AF615939

SU 0082840 8
621-522/FAW

HYDRAULIC SERVOMECHANISMS AND THEIR APPLICATIONS

J. R. FAWCETT, B.Sc.,M.I.Mech.E.

TRADE AND TECHNICAL PRESS LTD,.
MORDEN, SURREY, ENGLAND.

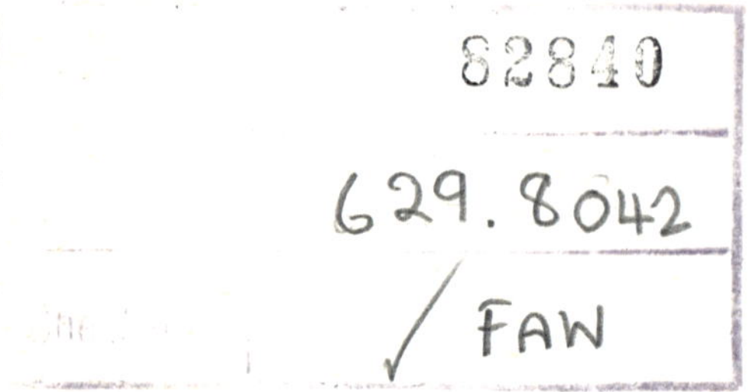

85461 026 X

PRINTED AND PUBLISHED BY TRADE AND TECHNICAL PRESS LTD.

CONTENTS

An Introduction

There are many books on hydraulic control systems addressed to the student or specialist, but these inevitably have a mathematical bias which the average engineer and non-specialist find difficult to follow. This book is intended to make the subject intelligible to the layman but this does not imply that a specialist will not still be needed in implementing the more complicated systems. Books for further reading are listed at the end.

Most engineers have formed ideas on what is meant by a servo-mechanism. The word 'servo' is derived from the Greek word for slave. Like the word 'automation' it is sometimes misused and taken to mean any mechanism which magnifies the power of the controlling force. A servo-mechanism is, strictly speaking, an amplifying system with kinetic output. In other words there must be a signal from the output end if there has been any deviation from the desired position or value which is continually moving or varying. Taking one of the earliest examples – an engine governor – the effect of the action initiated by the governor in adjusting the steam valve is fed back to it as a change of engine speed. The signal is usually known as 'feedback' and a system where feedback is present a 'closed loop' one. A simple amplifying system is known as an 'open loop' one.

Analysis will often show, however, that the popularly held ideas are illogical and depend on custom and practice. If we consider the ordinary motor car, for example, no one would term the engine and transmission a servo-mechanisms for converting movements of the driver's foot to a proportional amount of power at the wheels, yet servo-brakes are commonly referred to and cause no misunderstanding.

Or consider a digging machine with power steering; digging is controlled by the driver opening and closing valves in sympathy with what he can see is happening and there is no relationship between valve position and bucket position. The power steering gear incorporates a mechanism whereby the wheels turn the exact amount determined by the steering wheel.

The first is obviously not a servo-mechanism. The second comes within what we will take as our definition of a servo – a device which incorporated some non-human link between cause and effect. This includes force intensifying mechanisms whether operated humanly or by some other agency and systems incorporating non-human feed-back.

The fact that a human is one of the finest of servo-mechanisms should never be overlooked and the principal reasons for replacing him by a mechanical contrivance are often because the application is too heavy, or requires continual attention for long periods.

The earliest servos were probably those fitted to ships' steering and operated by steam. They have been largely replaced by hydraulically operated ones. The steering gear of the ocean going ship is unusual because once a course is set the steering may not have to be altered for hours, or maybe days. With steam steering it was the practice to start and stop the engine so that power was only being consumed for short periods. The same obviously applies to hydraulic steering and provision must be made for unloading pumps for long periods, including the use of accumulators.

Most servo-mechanisms, until the advent of the electro-hydraulic servo-valve, were specially made for the job. They included ships' steering gears of many kinds, gunnery control, ships' stabilizers, speed control of prime movers, machine tool copiers, vehicle power steering, forging press controls,etc.

The advent of the electro-servo valve, developed principally for aero-space projects, and since developed in more robust forms for commercial use, has changed all this.

Traditional servos have usually operated at fairly low pressures and had plenty of power at the primary stage. Reliability was excellent as there were no critical parts or orifices. Electro-hydraulic servo valves, however, are inevitably more critical – the power at the first stage is necessarily very small and there must be orifices or small clearances through which the fluid is passing continuously, giving opportunities for dirt to filter out in the vital parts of the valve. Modern filtration techniques and designs ensure that if a blockage does occur it does not cause the valve to go 'hard over'. It is fortunate that the results of the experience gained with the exacting requirements of aero-space travel have become generally available.

There are two general types of servo-controlled systems which for convenience can be called 'static' and 'dynamic'; although in servo language the respective descriptions are position-controlled and velocity-controlled. In static systems the controlled parts attain a definite position which is held for a discreet length of time. The mechanism is then really a convenient method of magnifying a force and, incidentally, in applying the magnified force at a place remote from the actual control point, e.g. a steering gear.

Dynamic systems, on the other hand, – for example a governor – are continually changing and are subject to the usual hazards of such systems, namely, hunting, overshooting and general lack of stabillty. It is, of course, this type of system with which control theory is concerned.

Any such system can be regarded as similar to the case of a weight supported by a spring. Trying to move the weight up and down via the spring causes the weight to get out of step with the upper end of the spring and the weight movement is further complicated by any friction. In a hydraulic system the spring represents the elasticity of the fluid on the ram or motor. Friction is present both in the mechanical parts and the fluid itself. One of the difficulties of theoretical treatment is that the hardest job is to get the relevant data on which a completely new system can be based. Practical experience on similar systems, followed by trial and error, is likely to give the best results. This is confirmed by the fact that whilst most text books on applied mathematics include numerical examples of applications, whereas in those dealing with servo - mechanisims such applications are conspicuous by there absence.

Electro-hydraulic servo-valves can open up new fields and give a fresh approach to old ones. They are the only servo-valves commercially available, although it is possible to operate some types mechanically. It is obviously essential that potential users should be aware of where servo-valves can help them in their problems and they should always be willing to co-operate with the suppliers.

SERVO VERSUS TRADITIONAL SYSTEMS

For a system involving any degree of complication it is always worth considering the use of servo-valves. This is best illustrated by considering an actual case, namely, one which concerns a number of unidirectionally loaded rams which have to be stopped at a number of fairly close fixed intervals, in both directions. To give a quick operation, without shock, the rams must be accelerated and decelerated smoothly but move at maximum speed for as long as possible.

The traditional method would involve three micro-switches at each stop, one to introduce a braking valve in the up direction, one to do the same in the opposite direction and the third to actually stop the ram. If the space is limited, as in this case, the ram movement would have to be taken by mechanical means to where the switches would be accessible. As the specification calls forf stopping at any selected station and not at each station in sequence, at least one switch to stop the ram would be needed at each station but it might be possible to arrange the circuit so that the braking switches were not duplicated.

With an electro-hydraulic servo-valve, however, it is only necessary to compare electrically a number of fixed resistances, with variable resistance operated by the ram. The appropriate fixed resistances are selected by the operator and the servo-valve opened by applying an electric current. The fixed and variable resistances are continually compared in a suitable transducer and when the two approach each other the current to the servo-valve is gradually reduced and finally the valve closes. If the ram fell, due to leakage, the servo-valve would automatically open slightly to counteract it.

Steering gears have already been mentioned and with the availability of gyroscopic devices a natural extension of steering gear techniques has been to the operation of stabilizer fins which reduce the roll of ships considerably and are now considered essential on passenger ships of all sizes. On larger aircraft similar principles have been applied to several flight controls.

Machine tools are using electro-hydraulic servo-valves increasingly. On numerically controlled contouring machines they are now the accepted means – in conjunction with hydraulic motors – of table position control. They are also used for the more complicated copying jobs, especially where contours change rapidly.

Endurance tests of all kinds are carried out on rigs controlled by electro-hydraulic servo-valves which can be driven from oscillators or operated to predetermined programmes recorded on magnetic tape.

Although most servo valves are of the modulating type there are some applications where a sensitive digital or 'bang-bang' valve is preferable, but these are exceptional.

Servo-Amplifiers and Steering Gears

In an open-loop system an input command or signal is fed via a controlled amplifier to an actuator providing an output against a given load - Fig. 1. The system, in effect, works purely as a power amplifier, with the output likely to be affected by variations in load and also variations in system parameters. Idealised characteristics are defined

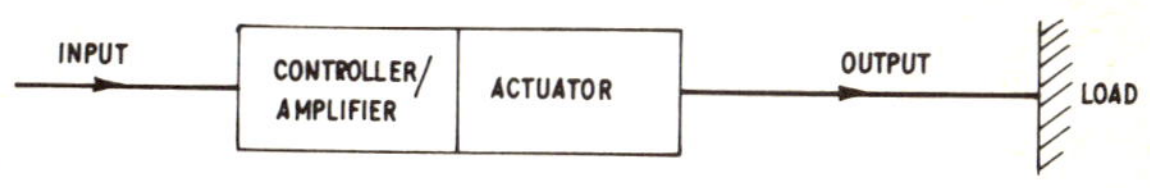

Fig. 1

in Fig. 2, where the controller is a manually (or mechanically) operated directional control valve providing position control of a linear actuator. Displacement of the valve from its neutral position will cause the actuator to move at constant velocity during the open time of the valve. Such a system is suitable for simple applications only which do not require accurate position control.

This latter limitation can be overcome by the addition of mechanical feedback, when the system becomes a simple servo-mechanism rather

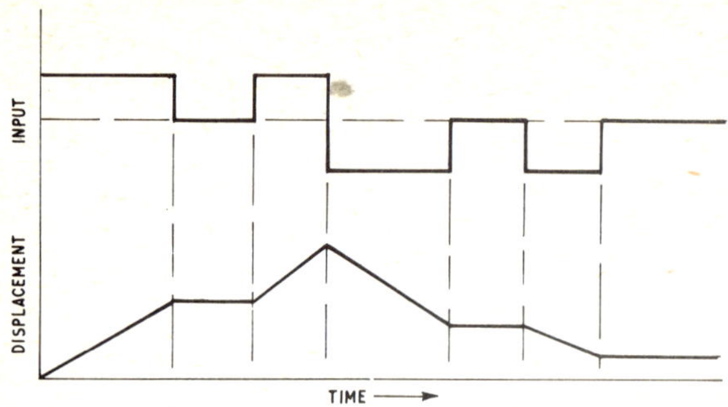

Fig. 2

than a pure amplifier. Basis of such a mechanism is shown in Fig.3 where input, control valve and actuator are interconnected mechanically by a differential lever. Movement of the input lever results in a corresponding movement of the spool of the directional control valve, by virtue of the differential lever pivoting about C, admitting fluid to one end of the actuator cylinder. Piston movement thus follows at constant velocity, with the differential lever now pivoting about A.

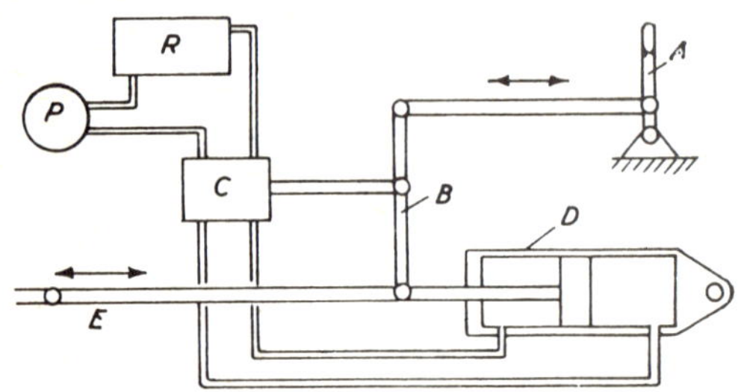

Fig. 3

A simple servo mechanism with mechanical feed-back. A - Input; B - Differential lever; C - Servo valve; D - Actuator; E - Output; P - Pump; R - Reservoir or tank.

When the piston has moved to a position corresponding to the input movement, movement of the differential lever has closed the control valve and so motion ceases. In other words, the differential lever provides a means of establishing a predetermined relationship between input movement and output movement by virtue of the fact that when this predetermined relation exists the differential lever will take up a position so that the control valve will be closed; and under any other condition the control valve will be open to produce self-correcting or follow-up action, characteristic of a true servo system. The scheme of dependence of one element on another forms a control loop, providing a continuous servo system. In practice, it is not necessary that the control valve and actuator be separate, and the two can readily be combined as one compact unit, as shown in Fig. 4.

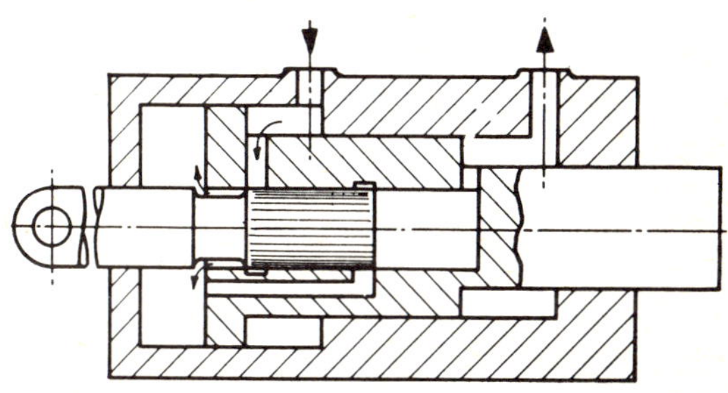

Fig. 4

FLOATING LEVER MECHANISMS

A typical floating lever system based on traditional steam practice is shown in Fig. 5. It consists of a cylinder with slide valve incorporated with the piston rod and valve spindle connected by the floating lever. Any movement of the floating lever opens the valve which automatically closes as the piston takes up the required position. It will be noticed that the lever has no fixed pivot, hence the term floating'.

This mechanism was invented when all slide valves overlapped' (see Fig. 6) as any other arrangement would have then been unthinkable both because of the waste of power with an overlapped valve and the difficulty of making zero-lapped valves with sufficient accuracy. It

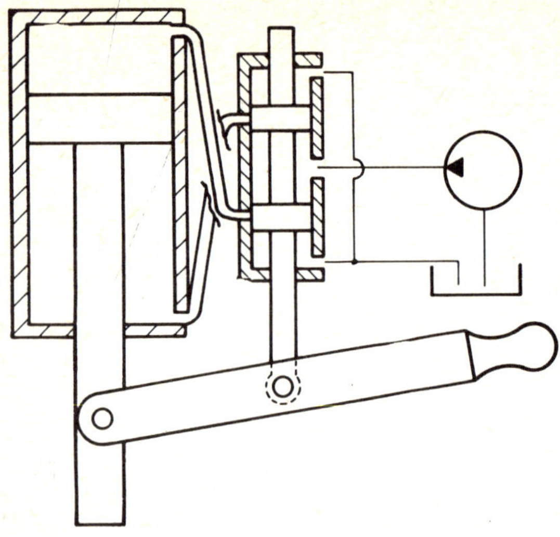

Fig. 5
Simple slide valve servo with position of piston controlled by hand lever.

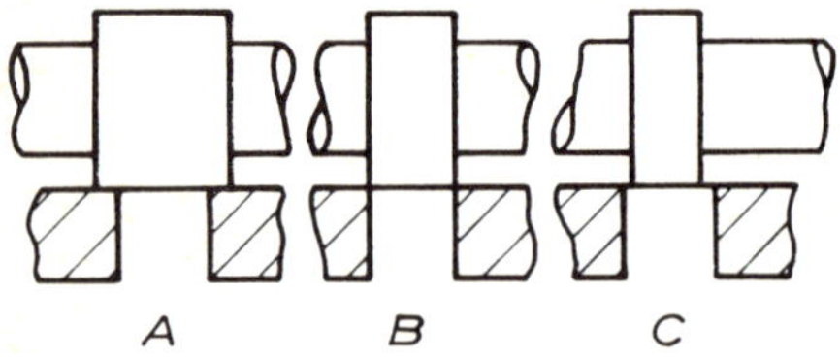

Fig. 6
A - Overlap. B - Zero lap. C - Underlap.

was not until many years later, when powered automobile steering was being developed, that the advantages of 'underlap' were realised for some hydraulic valves. Not only may underlap increase sensitivity, but if the amount can be made large enough without effecting the performance of the machine, it allows the pump output to flow back to the tank with relatively small back pressure when in the neutral position.

With automobile steering systems a pump which is large enough when the engine is ticking over – as it may be when turning the wheels during parking – has far too much output when running at normal speeds, and this is one way of dealing with the excess, although in practice it is used in conjunction with special relief valves.

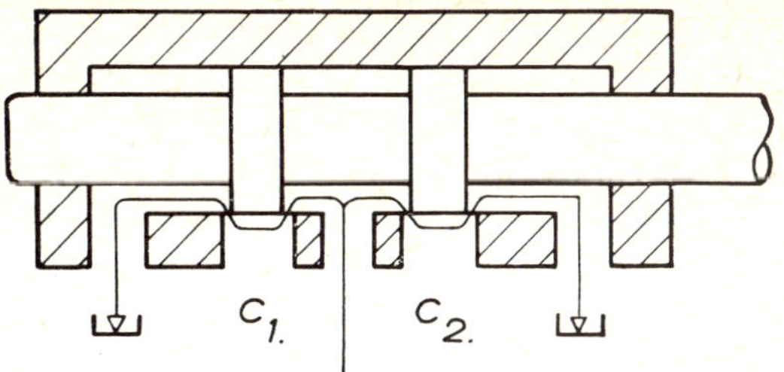

Fig. 7

Underlapped valve showing continuous flow in neutral position. Resistances to flow at C1 and C2 are equal.

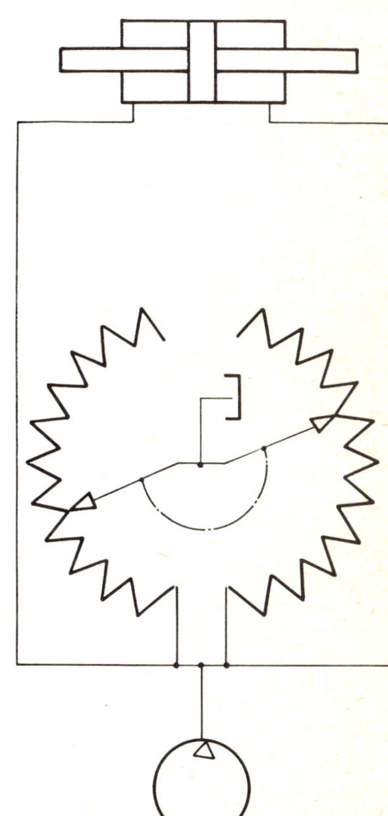

Fig. 8

Underlapped two-land valve showing that the effect is that of two coupled hydraulic potentiometers.

An underlapped valve (Fig. 7) can be likened to an electrical circuit (Fig. 8) with two variable resistances. With the valve in the neutral position the pressure drops are equal and the forces on both sides of the piston balance. Any movement disturbs the balance and the piston tends to move under the influence of the pressure difference. With an overlapped valve, it must be moved the full width of lap before anything happens. The ability to work with zero or underlap is one of the great advantages of hydraulic valves when precision and speedy response are required. This is not shared by other types; in fact with an overlapped valve there is a definite 'dead' zone.

When extreme accuracy is essential, as in copiers, movements of as little as .0001 in. (.0025mm) can cause a significant pressure change.

INCORPORATION OF 'FEEL'

With mechanism shown in Fig. 9 the operator has no idea of the force being exerted and this would only be limited by the available pressure. This can be a serious disadvantage as it may be possible to unwittingly cause damage or take an undesirable action.

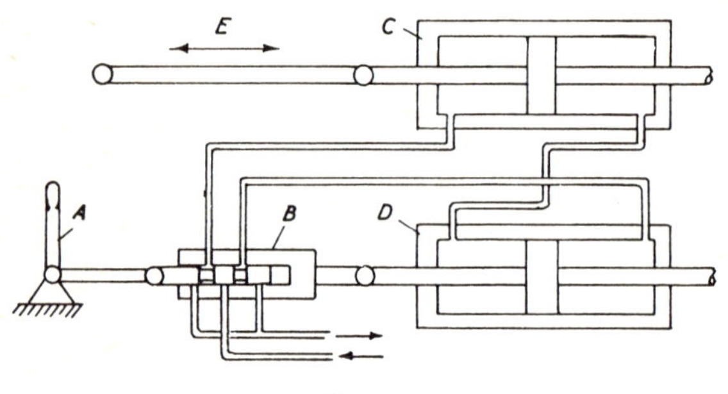

Fig. 9

A simple servo mechanism with hydraulic feed-back. A - Input; B - Servo valve; C - Actuator; D - Auxiliary cylinder; E - Output.

A method of introducing a reaction directly from the load itself is shown in Fig. 10. Instead of being applied directly to the ram, the load is connected to an extension of the floating lever, the dimensions being adjusted to give the proportion of the load which the operation must exert.

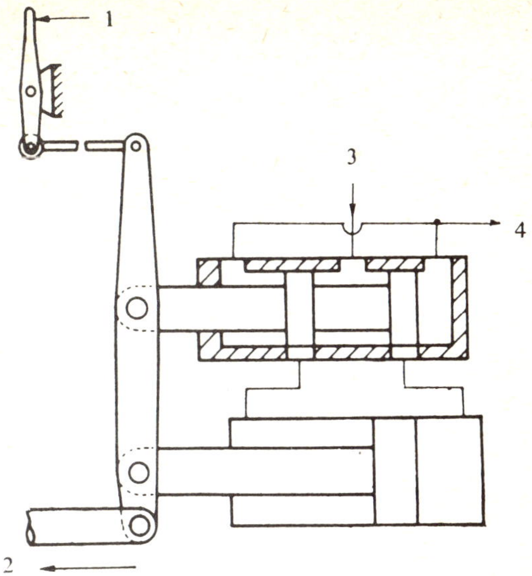

Fig. 10

Hydraulic servo with mechanical reaction or feel.
1. Input. 2. Output. 3. Pump. 4. Exhaust.

A neat method of providing 'feel' which has been applied to vehicle and aircraft controls is shown in Fig. 11 applied to one direction only. A small subsidiary piston is connected to the pressure in the head end of the cylinder and produces a permanent reaction on the operator's lever in direct proportion to the main load. The operator is therefore continuously aware of the force he is controlling. Sudden shock loads may, however, be absorbed without being transmitted to him.

LEVERLESS SERVO AMPLIFIERS

The mechanisms so far described all embody levers which are basically undesirable for many applications because of the space they occupy and the adverse effect of backlash which it develops in use.

Considerable interest attaches therefore to the type shown in Fig. 12 which has the valve incorporated in the main output member. Provided the pressure is sufficient to overcome the load this will follow faithfully the position of the input plunger. Similar units, but having underlapped valve spools, are very suitable for the first stage of electro hydraulic valves and will be described later.

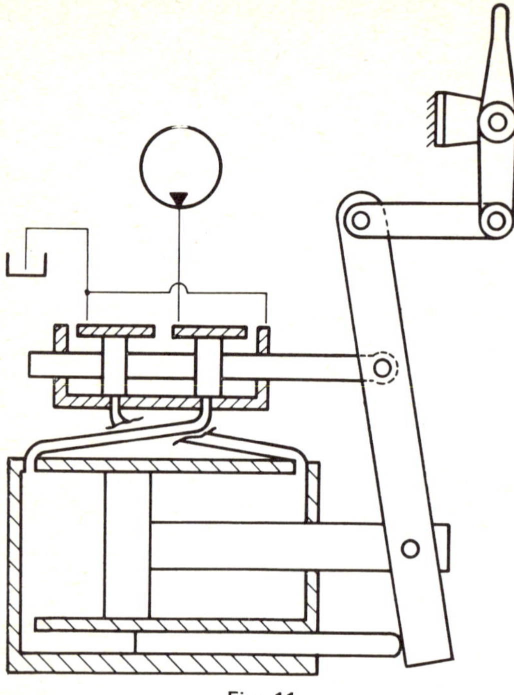

Fig. 11
Link servo with 'feel' cylinder. This principle is adopted when the user must be aware of the force being exerted at the output.

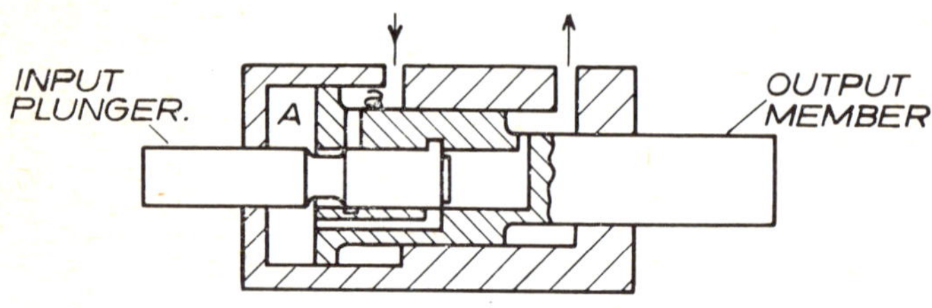

Fig. 12
Leverless servo where the valve is incorporated in the operating parts. The output member follows the movement of the input plunger exactly.

Moving the valve plunger admits or exhausts oil to the appropriate side of the piston until the neutral position is restored. Movement of the input plunger to the right causes pressure oil to be admitted to the annulus A, which being of larger area than annulus 'a', forces the piston to the right. Moving the plunger to the left allows oil to exhaust from A, so that the pressure in 'a' can take control until the valve again closes. For practical reasons the available stroke is usually short.

TWO-STAGE VALVES

Where the primary force is small it may be necessary to use a two-stage valve, and this is almost invariably so in the electrically operated valves described later. Fig. 13 shows an arrangement with two spool valves. The primary valve has underlap for greater sensitivity whilst the secondary, or main valve, has centralising springs which ensure that its movement is proportional to the signal from the primary valve. Without the springs the main valve would shuttle hard over with every change. One thing may be mentioned here – valves with underlap which remain in the neutral position for any length of time are particularly susceptible to dirt lock as the fine orifice acts as a very efficient filter. In practice, great care has to be taken to prevent this happening.

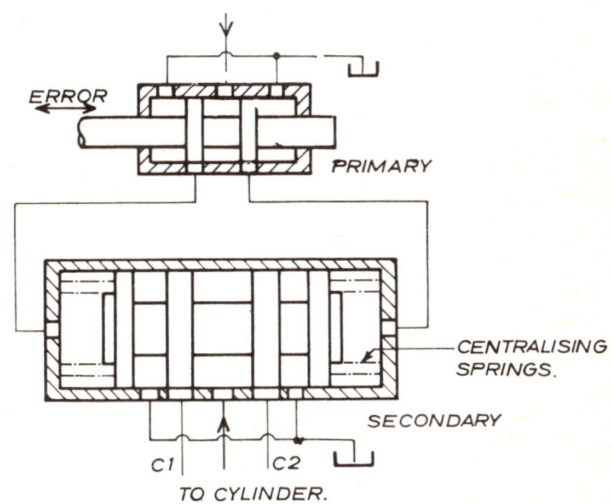

Fig. 13

Two stage spool valves. The primary valve is made as small and as sensitive as possible. The centralising springs ensure that the movement of the secondary spool is linear with that of the primary spool.

TWO-STAGE VALVE ARRANGED AS CONTROLLER.

The requirements of controllers are dealt with in some detail in Chapter 9. The layout in Fig. 14 does, however, demonstrate how the same problem has been solved hydraulically. This has the characteristic of responding to sudden changes of the controlling valve ('error') and then restoring the valve to its original state. The method is simple in principle but when included in an actual mechanism tends to look so complicated that it is very difficult to follow.

Interposed between the governor or other sensing device and the normal control is placed a dashpot and the thrust operating the main valve is taken through the dashpot to a centring spring, i.e. whether pushed or pulled it returns to the same length. Any momentary movement of the governor is taken via the dashpot to the spring which stretches or compresses to allow the control to operate in the usual way. If the change is permanent the dashpot will now tend to react to the force on it and will continue to do so until the spring is restored to the new position.

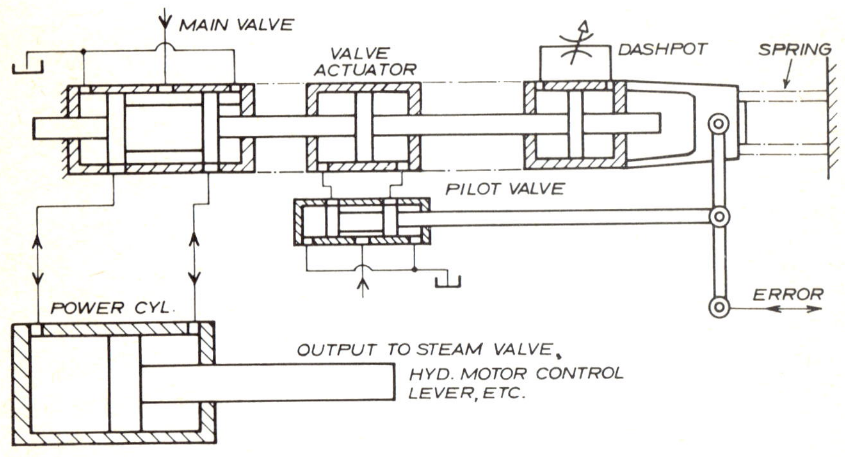

Fig.14

Two stage valve arranged as controller, with provision for reset. Errors are signalled from the governor or other error detector. Restoring action is taken and the preset value gradually restored. Sensitivity is governed by the dashpot setting. The parts are shown disconnected.

STEERING GEARS

Power assisted steering gears for cars and mechanical handling vehicles are made in large quantities and are good examples of servo-mechanisms. They may also be the cheapest kinds available and are a convenient way of applying power assistance to movements which do not lend themselves to control by the ordinary hydraulic cylinder and four-way valve.

A steering booster designed for slow moving vehicles and also for general use is shown in Fig. 15. It is intended for fitting, with little modification, to a standard steering linkage, the main modification being to the drag link from the steering box and the steering arm.

The drag link operates an open centre four-way valve within the booster head. The valve is spring centred and turning the steering wheel causes the drag link ball stud to push the valve spool one way or the other and so direct oil to the appropriate end of the cylinder. As soon as the correct position is attained the valve centres and the pump output passes back to the tank. Working pressure is 400 to 600 psi (28 to 42 Kg/cm^2). A practical detail is that the valve is by-passed should the vehicle have to be steered with the engine stopped.

This steering booster can also be used for applying a force of a few hundred lbs over a distance of a few inches and is very suitable where heavy controls have to be adjusted. The booster is interposed between a solid support and the part to be moved (Fig. 16) and the lever which would otherwise be used to move the load direct takes the place of the drag link. Pressure on the lever opens the valve in the booster in the appropriate direction, causing it to extend or retract as required, with a minimum of effort on the part of the operator.

When power assisted steering is applied to fast moving vehicles it is very desirable that the driver or pilot should be aware of the force which the mechanism is exerting. The mechanism is therefore arranged so that a proportion of the load is applied to the driver's controls, so providing 'feel'.

For fast moving vehicles, commercial practice is to incorporate the power assistance in the steering box and an example is shown in Fig. 17. It will be noticed that 'feel' is incorporated. The steering gear is based on the Marles cam and double roller steering which can operate the rocker shaft direct should the power fail.

Servo control is provided by a set of four valves J and H, one pair for each cylinder. When no power is required oil flows into each cylinder from valve G, out by the second pipe to valve J and back to the tank, all valves being open and allowing free circulation.

Reaction to operate the servo is provided by the force between the spur gears O which tend to move the shaft E sideways. Shaft E is

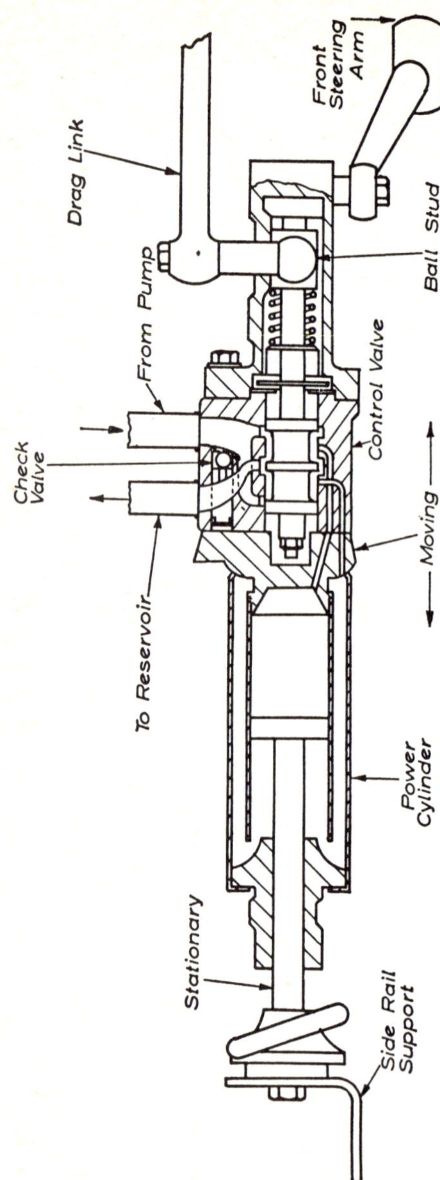

Fig. 15

Booster for power assisted steering, incorporating double acting cylinder and valve in head.

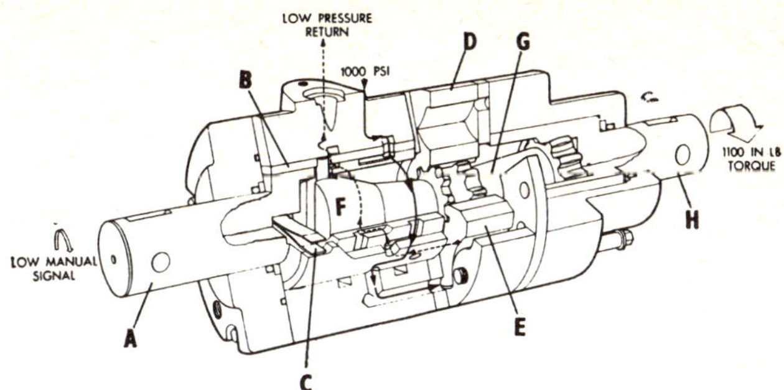

Fig. 16
Steering booster with internal gear. A - Input shaft. B - Valve sleeves. C - Valve centralising spring. D - Hydraulic motor external gear. E - Hydraulic motor internal gear. F - Input connecting shaft. G - Output connecting shaft. H - Output shaft.

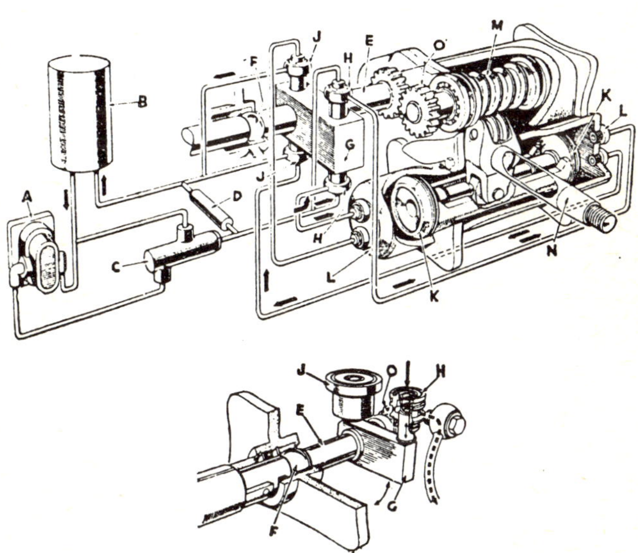

Fig. 17
Steering box with power assistance (Marles).

supported at the other end by the spherical joint F and is turned by the steering wheel and runs freely in block G. This presses on the appropriate reaction valve J to close it and so blocks the exhaust from that cylinder. At the same time the valve H, controlling the entry of oil to the other cylinder, is closed.

With this arrangement the reaction valves J resist the movement of the block and so cause a reaction in the steering proportional to the pressure in the cylinder to give the driver 'feel'. Any sudden torque on the rocker shaft, such as that which may arise from a wheel striking an obstacle, will cause the valves to operate in such a direction that pressure is built up to oppose the torque. Unless the speed is very low, no reaction is transmitted to the steering wheel. The self-centring action of the steering is not destroyed by this gear and if the vehicle tends to self-centre without the gear it will do so with it.

Valve C is the bypass flow control valve previously described. To provide for steering when the pump is inoperative for any reason, a bypass valve D allows oil to circulate as the gear operates. There is a certain amount of backlash as the valves move for the full extent of their travel (about .020in .5mm) before the gears transmit the torque.

With the 'Saginaw' steering gear the operating valve is interposed in the shaft from the steering column and forms a self-contained unit with the steering box. The gearbox shaft is screwed at the end into a nut having cut in it rack teeth which engage with the gear on the drop arm shaft. With a normal unassisted steering box, turning the shaft causes the nut to move along the screw and turn the drop arm shaft, but in the assisted pattern, the reaction causes the steering column shaft to move endwise slightly and so actuate the valve.

The 'feel' or reaction is provided by a number of plungers which are connected to the pump pressure and which bear against flanges on the valve spool. If the resistance to steering is high, as when steering at low speed or turning the wheels when the vehicle is stationary, the pressure builds up to overcome the resistance and the reaction plungers make it proportionately difficult for the driver to move the valve spool. The driver is therefore made conscious of the effort being exerted by the servo, even if he is in fact only called upon to apply about one quarter of the effort which would otherwise be necessary.

As in many other servo-mechanisms, this design tends to insulate the initiating end from shocks occurring at the output end. With the Saginaw gear severe shocks at the road wheels are absorbed by the booster piston and are not felt by the driver. The shock causes the valve to shut off the appropriate end of the cylinder and the relatively incompressible oil prevents movement of the steering.

A rather different principle has been adopted in the Danfoss steering gear, but the motion is rotary rather than axial. The steering wheel shaft has at its lower end one element of the rotary valve. Turning this admits oil under pressure to the special internal gear motor which rotates the output shaft and at the same time the other element of the valve follows up and shuts off the pressure supply when the new position is reached.

In an ordinary internal gear pump the inner gear is attached to the driving shaft and the outer internally toothed gear is arranged so that it can rotate and does, in fact, revolve at a fraction of the shaft speed, which depends on the number of teeth; e.g. with six teeth on the inner gear and seven on the outer, the ratio is 6:1. If the outer gear is fixed and the inner gear is revolved, it will be found that the centre of the inner gear now runs eccentrically. Also, for one revolution of the inner gear the centre of the gear revolves round its axis six times for each revolution of the gear itself.

The practical effect of this is that the displacement per revolution of the inner gear is six times what it would be if the outer gear were free to revolve, so that the power when used as a motor is multiplied six times.

Referring to Fig. 18 it will be noticed that shafts are arranged with a spherical end and rounded teeth so that they can follow the eccentric motion of the inner gear. The input drive can be comparatively light, but the output must be capable of dealing with the full torque.

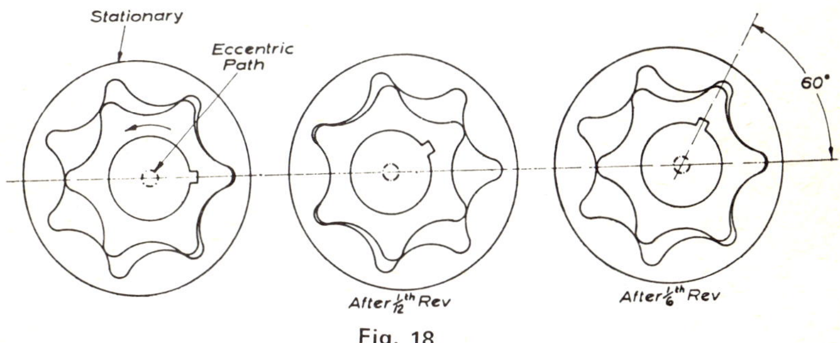

Fig. 18

Successive positions of the inner gear, showing how the eccentric axis revolves six times for one revolution of the gear itself.

The valve sleeve forms an open-centre four-way valve, and is normally kept in the open position by a spring. Any movement of the input shaft causes the valve to divert the oil to the appropriate side of the motor gears.

In the event of power failure the input shaft can drive the output shaft direct, the pins engaging with the spherical end of the shaft taking the drive.

Two of the basic types of ships' steering gear are shown in Figs.19 and 20. The Rapson slide gear (Fig.19) comprises, basically, one or two pairs of opposed hydraulic cylinders, the rams of each pair being coupled together to form a double bearing in which pivots a ram cross head which is free to slide along the circular tiller arm on which it is mounted. A typical rotary vane unit is shown in diagrammatic form in Fig.20.

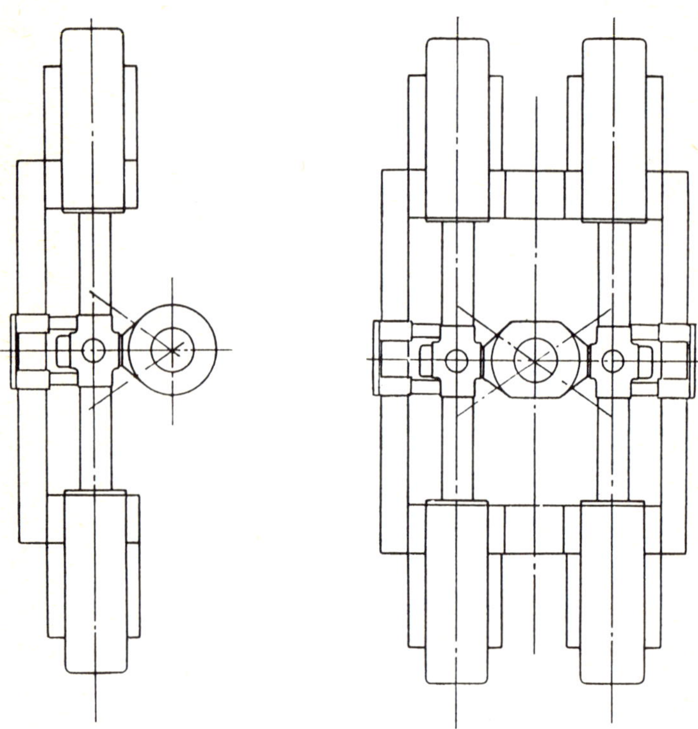

Rapson slide gear.

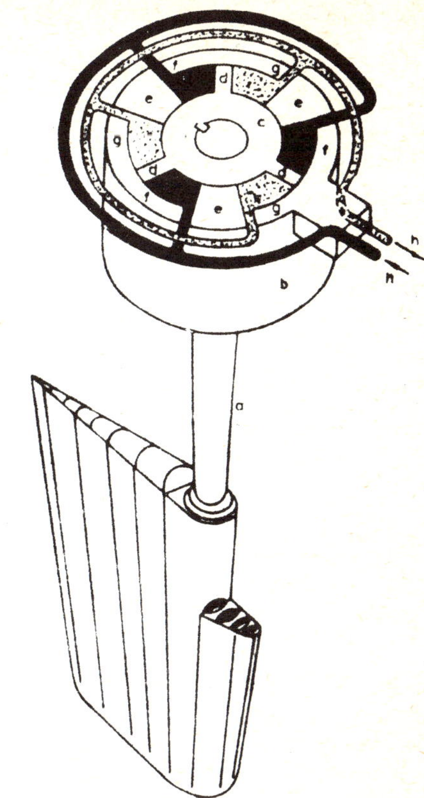

Fig. 20
Rotary vane unit.

Although very different in construction, all these devices are closed loop systems with the feed-back incorporated in the valve gear. As the command is given by turning the steering wheel so the 'error' causes the valve to open and when the turning stops the servo continues to operate for a short time until there is sufficient error in the reverse direction to close the valve. With mechanisms such as these the rate of operation is comparatively slow and no account has to be taken of tendencies to hunting or instability as occur in fast moving mechanisms.

SHIPS' STEERING GEARS

A ship's steering gear arrangement which owes much to the original steam powered gears is shown in Fig. 21. The floating lever is supported

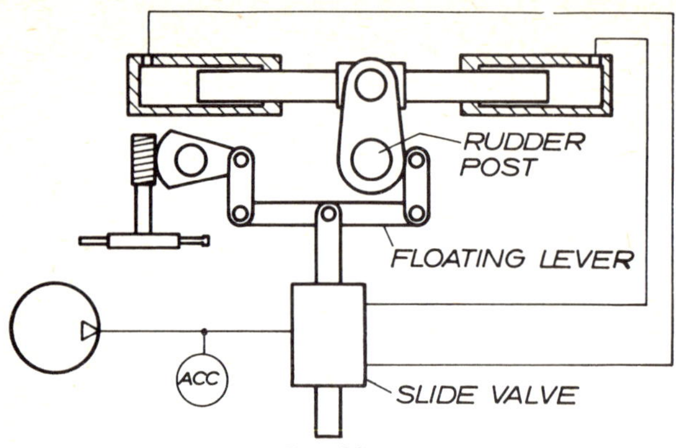

Fig. 21
Ship's steering gear powered by electrically driven pump. Normally the steering wheel is some distance away and connection is made by telemotor (see text).

at the centre by the slide valve spindle, the ends being connected to the steering wheel gearing and the other to the rudder posts. Until electrical systems became universal, power would often be supplied by a small steam engine which drove the hydraulic pump. (Fig. 22) The engine would only run when the steering wheel was turned. With all-hydraulic

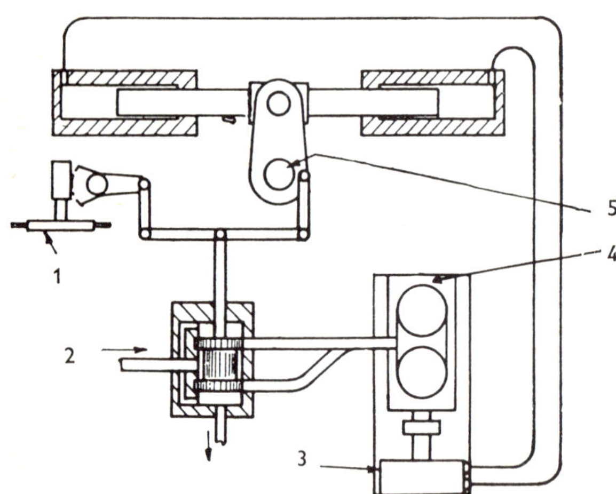

Fig. 22
Steam driven hydraulic steering gear with low steam consumption as the steam engine only runs when power is required. 1. Wheel. 2. Steam inlet. 3. Oil pump. 4. Steam inlet. 5. Rudder post.

systems various methods are used, some running the pump continuously, others only when required.

When, as is usual, the distance between the wheel and rudder is too great for direct mechanical operation, it was customary to link wheel and steering mechanisms by a telemotor. This is not a servo drive but a means of transmitting hydraulically the wheel movements to the rudder servo from the bridge – sometimes a considerable distance. The wheel is geared to two master pistons one of which moves in and the other out when the wheel is turned. At the distant end two slave pistons operate a slide which is connected to the floating lever in place of the direct drive shown. With the increased reliability of electrical transmitters these are tending to supersede the telemotor.

MODERN SHIPS' STEERING GEARS

An example of the use of an electric feed-back system is provided by the Vosper system, the hydraulic circuit for which is shown in Fig. 23. The constant delivery oil pump runs continuously. A bleed-off flow control valve with coarse and fine adjustment determines the effective pump delivery and therefore the speed at which the rudder turns. The open centre valves unload the pump when the rudder is stationary.

There are three 3-position open centre control valves in parallel and these are connected to the pair of double acting cylinders through operated check valves. The master and standby control valves are solenoid operated from the electrical system. Potentiometers at the wheel and rudder head generate signals which are handled by a fully transistorised switching unit.

If the master valve fails, it is automatically isolated and the standby valve substituted. Should the electrical system fail, then the third valve can be either telemotor or manually controlled.

The system is suitable for vessels up to 25,000 tons, 1500 psi (105 Kg/cm^2) being used for torques up to 100 ton-ft and 3000 (210 Kg/cm^2) for higher torques, so reducing the ram size.

The electrical controls, including the cables, are duplicated, each set being independent of the other. Normally the rudders respond exactly to the wheel movement, but push button controls which move the rudder whilst depressed are also provided. The rudder movement is then indicated at the bridge by an electrical instrument. Provision is made for the gyro compass or auto pilot signal to be fed into the system.

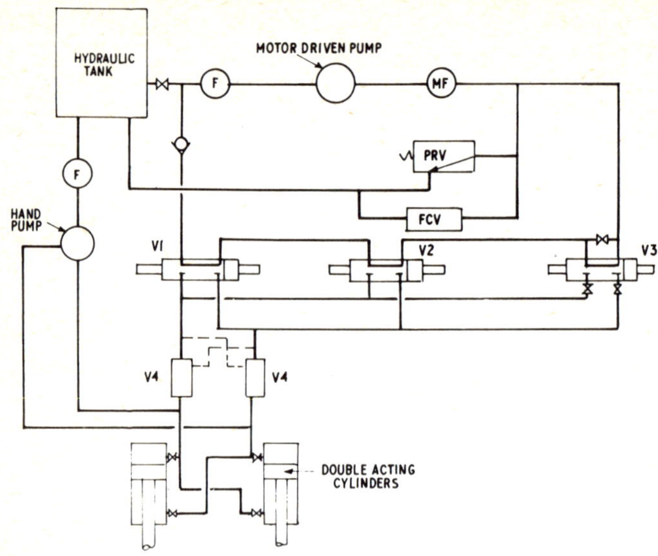

Fig. 23

Typical hydraulic circuit for Vosper steering gear.

F - filter
MF - micro-filter
PRV - pressure relief valve
FCV - flow control valve
V1 - manual power control valve
V2 - secondary control valve
V3 - master control valve
V4 - pilot operated check valve

Control of the variable delivery pumps which are usually pteferred for these installations is by a floating lever system.

Self-contained servos on the pumps, usually similar in principle to the leverless type shown in Fig. 12, require little mechanical effort to operate and the floating lever mechanism can be housed in a self-contained box lubricated for life.

Jet Controlling Devices

Although not usually hydraulic, jet controlling devices of the type usually known as Process Controllers, exhibit many of the characteristics of closed loop hydraulic and electro-hydraulic servo-mechanisms. Because of their comparatively slow action they provide a very useful introductory study for those not acquainted with the practical problems of closed loop control - i.e. where the results of the controlling action are continually being fed back (hence "feedback") to the controller.

Process controllers are, in general, concerned in maintaining a physical quantity such as liquid level, flow, pressure or temperature as close to the desired value as possible. The actual value is continuously compared with the desired value and signals are sent out to the regulating unit – usually a modulating valve – to take the necessary corrective action. These modulating valves open in proportion to the strength of the signal sent to them. This is frequently an air signal which varies between 3 and 15 psi. The valve is fully closed when the signal pressure is 3 psi and fully open when it is 15 psi.

In most practical applications of closed loop servo-mechanisms the response to the signals is affected by inertia, friction, elasticity and delay in detection of the variable. With process controllers inertia and slow signal response are often the biggest factors and they are therefore designed primarily to deal with these. The same system will, however, deal with other factors.

METHODS OF CONTROL

It is a fundamental consideration of all automatic controls that they are 'error operated', in other words, no action can be taken until the output has deviated from its desired value. If the system is made too sensitive, then hunting may occur. On the other hand it may be possible to anticipate the effect of a condition which would otherwise cause an error. For instance, the speed of a steam turbine driving a generator is governed by a signal derived from a rise or fall of speed. But as the amount of steam required is tied closely to the electrical load it is possible to determine the load electrically and use this signal to operate the steam valve before the turbine speed has time to alter significantly.

Proportional control

All controllers incorporate this basic method, wherein the controlled value varies with the load. A simple example is the ordinary Watt governor, where an increase of load reduces the speed and vice versa.

Refinements in controllers are concerned with modifying or eliminating the effects of load on the output value.

Integral control

This is superimposed on the basic control and is said to integrate the error with respect to time. (Described later)

Derivative control

This control action is the derivative of the error in respect to time. Usually the signals are mixed in the controller and presented as a combined signal. As the last two methods involve time the use of internal orifices and capacities in conjunction with compressed air is invaluable.

JET AND FLAPPER MECHANISM

The basis of most process controller detecting mechanisms is one of the most sensitive fluid devices – the jet and flapper system (Fig.24) responds to very small forces and is capable of responding to movements of 0.00001 or less. It forms the basis of air gauging systems, where the part to be measured replaces or actuates the flapper of most pneumatic process controllers and positioning devices. It is widely used in the first stage of electro-hydraulic servo-valves and has also been incorporated in copiers. As a jet is larger for a given flow than the clearance between the spool and body of a slide valve it is less likely to become blocked, especially by long needle-like particles.

The sensitivity of such a device depends to some extent on the shape of the end of the nozzle. With a sharp nozzle the back pressure varies as the distance x, provided that x is greater than d/5. With a flat-ended nozzle the back pressure tends to vary as x^3 and the force on the flapper due to the airstream is considerably greater.

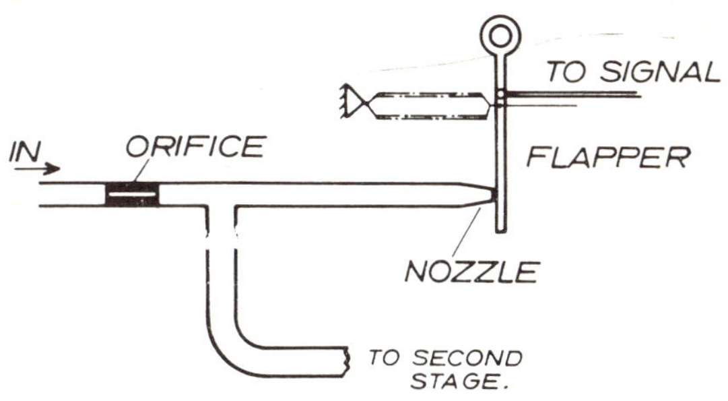

Fig.24

Jet and Flapper. If the flow from the nozzle is impeded, the back pressure builds up and can be used as movement detection device.

PNEUMATIC CONTROLLER MECHANISMS

In process controllers a nozzle and flapper system is associated with a servo-mechanism. Any change which results in movement of the flapper causes the back pressure to rise or fall and this in turn causes a pressure-sensitive element to initiate appropriate action. The air jet thus acts as a detector only, the associated mechanism usually being arranged so that it tends to keep the nozzle/flapper relationship constant.

The scheme shown in Fig.25 is used for controlling pressure or flow by opening or closing a spring-loaded diaphragm operated valve in accordance with changes detected by the measuring element. This element operates the flapper, tending to close the jet when the detected force increases, thus causing the diaphragm to raise the ball valve in the relay and release some air from the process valve, thus closing it. As the alteration to the valve setting takes place the measuring element responds, opens the nozzle and so causes the relay either to balance the valve pressure or increase it. As the amount the valve opens is dependent solely on the pressure applied to it there will be a slight drift with the load or valve opening. This can be offset by providing variable air resistance, restoring the drift and varying the sensitivity of the controller to suit the system being controlled.

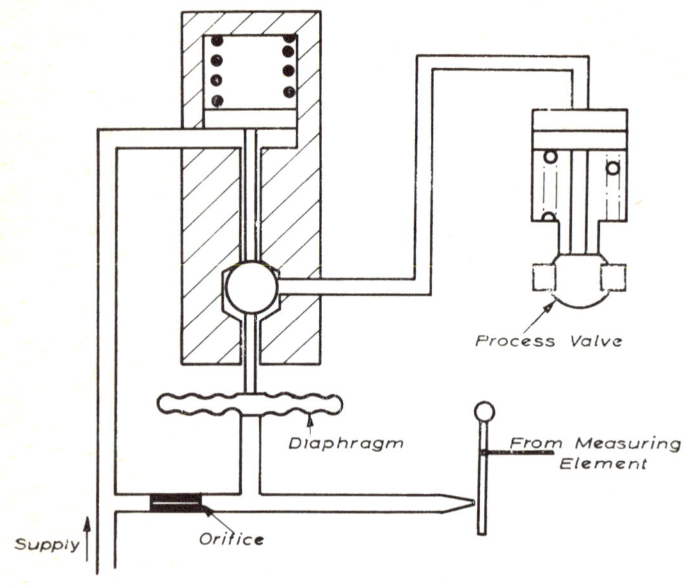

Fig.25

Simple pneumatic controller mechanism with diaphragm operated relay.

A floating lever mechanism is shown in Fig.26 The lever is pivoted at one end on a shaft attached to a flexible metallic bellows which extends in proportion to the applied pressure. Movement of the control varies the nozzle obstruction, causing the bellows to contract or expand, restoring the centre of the floating lever to its original position. The output pressure is thus in exact proportion to the control movement.

If a distant control requirement can be transferred directly to an equivalent pressure then it can be arranged to position a piston so than its position corresponds exactly with that of the direct control, Such a position controller is shown in Fig 27 where control pressure is applied to a metallic bellows connected to a floating lever passing between two nozzles. The far end of the lever is in contact with a cam or wedge attached to the piston.

Movement of the bellows causes one of the nozzles to be closed more than the other, thus upsetting the balance of pressure in the

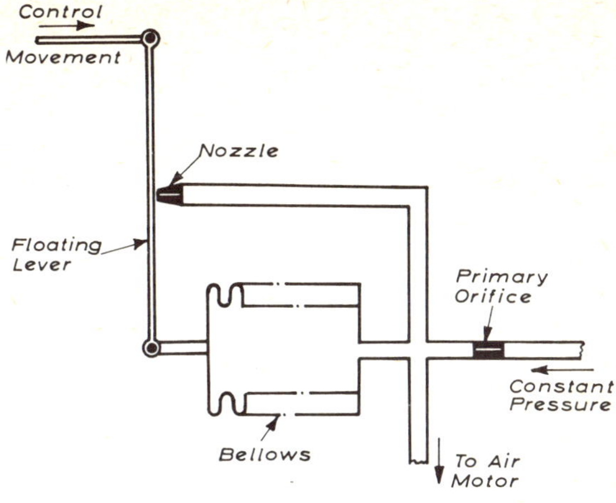

Fig. 26

Floating lever mechanism giving an output pressure in exact proportion to the control movement.

cylinder causing the piston to move until the cam has restored the floating lever to its original position. If an external force tends to move the piston, there is a slight movement which causes the wedge to move and pressure on the appropriate side of the piston is increased to compensate.

Where the load is small a single-acting spring-return piston may be sufficient. In this case the system of Fig. 28 could prove quite adequate.

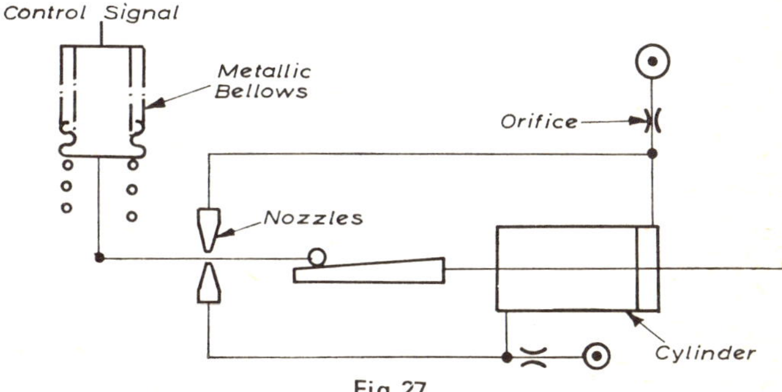

Fig. 27

Piston positioner which gives a precise piston position according to the pressure of an air hydraulic control signal acting on the bellows.

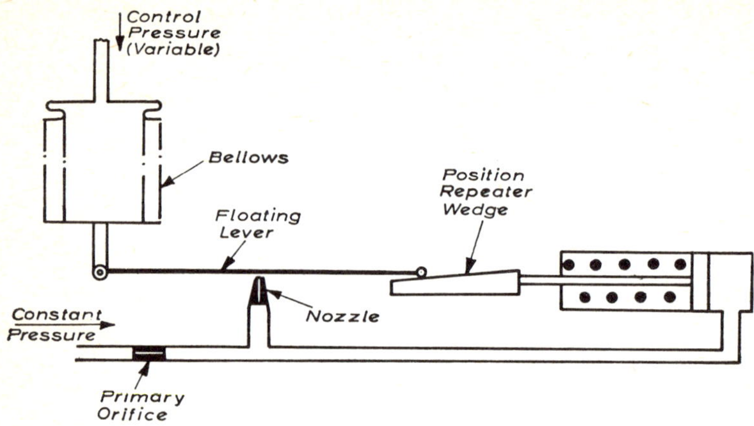

Fig.28
Piston positioner for light loads where a return spring can be used.

Jet Pipe Controllers

A system is shown in Fig.29 This is a jet pipe controller or pressure recovery method of control, giving large amplification of a small controller force. The jet is hinged and can be directed into one of two orifices, each orifice being connected to the end of a cylinder. Only a very small force is required to divert the hinged jet, changes of jet position tending to cause the piston to make a full stroke.

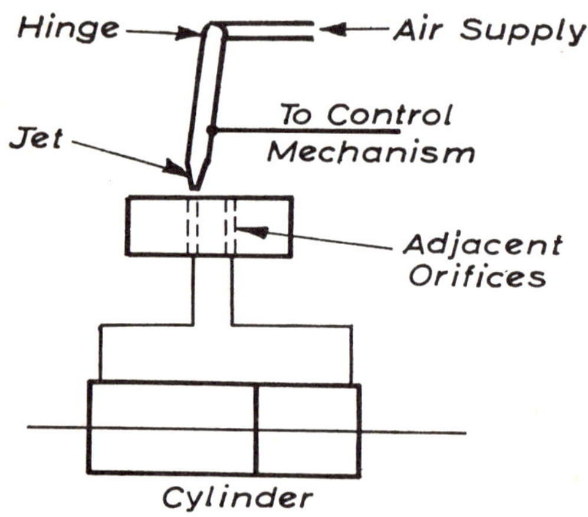

Fig.29
Jet pipe controller, giving large amplification of a small controller force.

Such a controller is often called the Askani regulator. Because of its inherent reliability it has been used for many years in heavy industry, where its characteristics also happen to suit the particular applications. More recently it has been used as the basis of what is claimed to be a most reliable electro-hydraulic servo- valve.

This controller has an intermittent action, rather than the continuous modulation associated with most servo-valves. As will be seen from the diagrammatical representation in Fig.30 the jet pipe J is pivoted on a horizontal axis at its upper end and oil at a constant pressure (usually about 100 psi) is pumped through orifice O. The controlling force, usually supplied by a pressure or temperature sensitive instrument is applied at F and the spring S gives adjustment. The plate D has two narrowly separated holes which communicate with a cylinder. When the nozzle discharges into either of the holes the kinetic energy in the jet is converted into pressure and conveyed to the appropriate end of the cylinder. In the neutral position the jet strikes the plate and is diverted. Normally the cylinder would operate a valve directly. As the piston is unpressurized and free to move in the neutral position, the controlled valve must be of the balanced type. By adding a subsidiary valve and cylinder it is possible to provide proportional control, but this has little practical application.

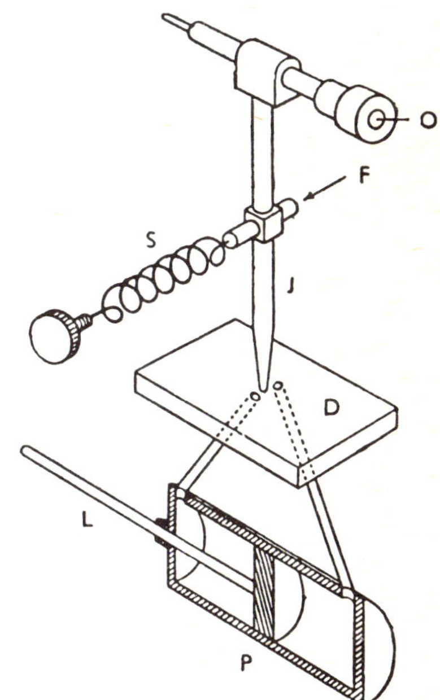

Fig. 30

Jet pipe controller, sometimes called Askani controller: J. Jet pipe; O. Oil Inlet; L. Link to valve; S. Setting spring; D. Distributor plate; F. Controlling force; P. Actuator piston.

Fluidic Devices

More recently fluidic elements have been employed as detection devices, with the advantages of high sensitivity and immunity from dust, moisture, vibration, etc. These normally take the form of back pressure sensing devices, or interruptible jets.

Integral plus Proportional Control

The main purpose of adding integral (or reset) control to proportional control is to correct the offset inherent with proportional control without making the control too sensitive as to cause hunting.

This involves the addition of another similar bellows to the original system Fig.31. If conditions are constant, the pressure in both bellows is the same and they balance each other. Any change of load causes the usual reaction on the proportional bellows and there is the corresponding offset in the liquid level. The proportional and integral bellows are connected through an adjustable restriction valve. As the pressure in the proportional bellows changes, there is a flow through the valve from one bellows to the other. If there is a permanent change of load, the pressure will equalise at a rate depending on the opening of the connecting valve. This has the effect of restoring the two bellows to their original position so that the control pressure is kept at its new value whilst the level is restored to normal. A similar effect would apply to, say, a speed control.

In practice, loads vary erratically and times of response may be appreciable. A compromise would be obtained by adjusting both the integral valve and the proportional band percentage until the conditions were as near stable as possible without instability.

It will be seen that the term 'integral' is connected with the rate of change of jet pressure – the greater the rate of change, the greater the pressure difference between the two bellows and also the greater the velocity through the connecting valve.

PROPORTIONAL PLUS DERIVATIVE CONTROL

Here the proportional element movement varies with the rate at which deviation changes. The time factor is obtained by inserting a resistance in the line between the nozzle and proportional bellows. The delay in the pressure build-up in the bellows has the effect of reducing the feed-back when deviations take place quickly but having little effect when they take place slowly. It therefore causes the control valve to open or close rather more than it would have done otherwise.

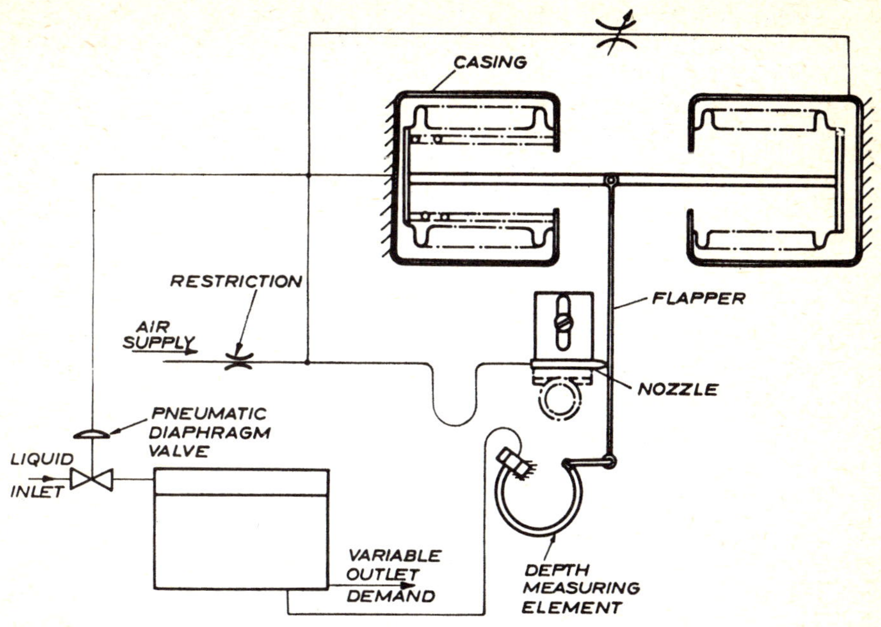

Fig.31

Proportional plus integral control. The addition of the second bellows, whilst not affecting the proportional effect which makes for stability, tends to restore the level to normal despite variations of demand.

Many commercial controllers have both integral and derivative in addition to proportional control and by adjusting them suitably practically all conditions can be catered for.

APPLICATION OF AUTOMATIC CONTROL

Although the pneumatic controller is primarily thought of as suitable for dealing with fairly slow changing variables, the principle incorporates are fundamental to all control problems.

With larger plants the additional and unavoidable delay in transmitting pneumatic signals over long lengths of pipe has turned attention to electronic controllers which can incorporate time delays of any length (3 seconds to 50 minutes) and these could be linked with electro-hydraulic servo-valves should the need arise.

The time delays are usually formed by resistances and capacitors whilst potentiometers give the various voltages. A manually operated potentiometer, for example, gives the desired value as a voltage and this is compared with the signal from the measuring element. Signals are amplified by transistors and the final output emerges as a 0–15mA DC current which is applied to a pneumatic or hydraulically operated control valve, often in conjunction with a hydraulic valve positioner.

EFFECTS OF CONTROL METHODS

With all methods of control there is an initial overshoot and the oscillations gradually die away (assuming the rate of the new demand is maintained). See also Fig. 32

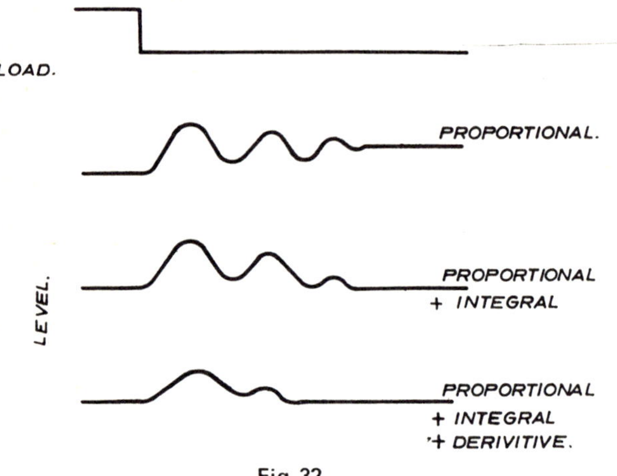

Fig. 32

Graphs to show behaviour of system after a sudden fall of load with various control methods.

Proportional control – gives a final value somewhat lower than the original.

Proportional plus integral control – gives, firstly, the same curve as before but the value gradually increases until it is approximately the same as the original. For this reason integral is often called 'reset'. The time it takes to do this depends on the integral control valve setting but would normally be as short as possible consistent with stability. If this method is used for speed control, the speed can only be an approximation of that specified and if, as on frequency controlled electric supplies, a definite number of revolutions per day is required, then the 'level' value will have to be altered from time to time to bring it into synchronisation with the frequency standard. This effect is only noticed by the public when the system is overloaded.

Proportional plus integral plus derivative control – This is exactly the same as before except that the addition of derivative control tends to slow down the proportion reaction and reduce the initial swing or kick through over correction.

One lesson which can be learned from the study of automatic process controllers is that control with an instantaneous response is not necessarily the best method of controlling and a more stable system can be obtained by having time lags consistent with the inertia of the system being controlled. See Fig. 33.

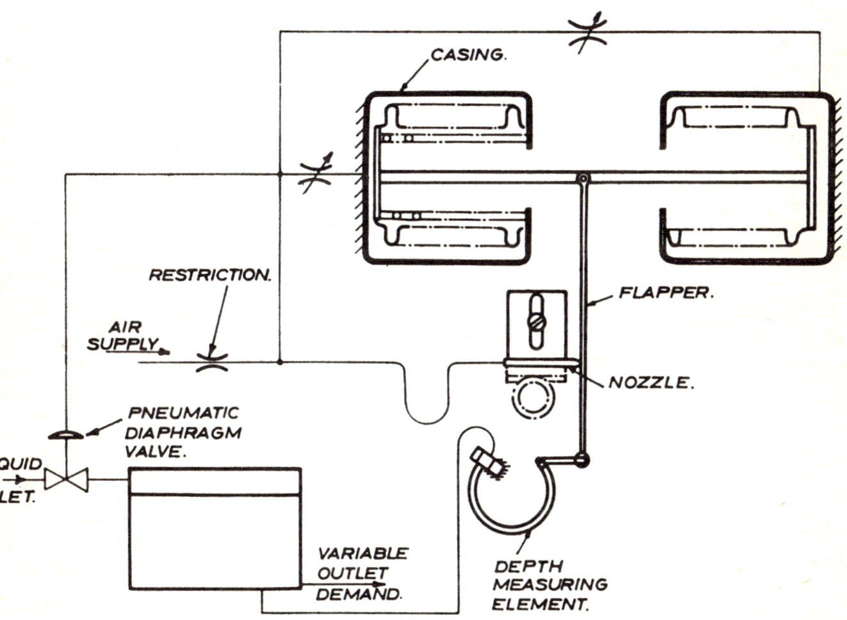

Fig.33

Proportional plus integral plus derivative control. This is similar to Fig.31, except for a regulating valve in the line to the proportional bellows which enables the rate of response to slowed down on systems with long time lags.

Pulse-length Modulation

Pulse-length modulation (PLM) is a method of digitally modulating bistable switching valves in a typical servo circuit – Fig. 34. In this circuit S1 and S2 are the two bistable valves, periodically switched from A to B in a symmetrical manner as long as no error signal exists. When an input signal is applied to the pulse-length modulator, the dwell times of the valves at positions A and B are varied differentially, thus providing differential flow to the actuator which moves in the appropriate direction. Motion ceases when the error signal falls to zero and the valves are again switching in a symmetrical manner.

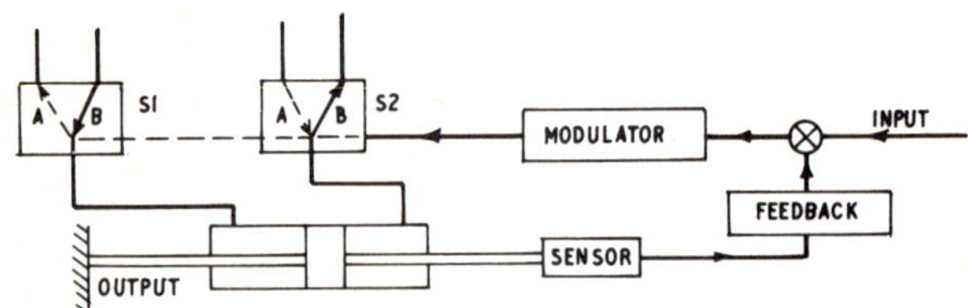

Fig. 34

This system of using bistable switching valves working as straightforward switches, as opposed to proportional valves, can be of considerable advantage in particular environments, especially when applied to pneumatic systems. It has been widely applied to missile servo systems.

Flapper Disc Valves

Flapper disc valves have come into prominence as the switching elements in differential pulse-length modulation (DPLM) systems. Such systems avoid the high quiescent power loss inherent in a straightforward PLM system and provide a closed-centre system adaptable to fluid valving techniques.

Such a valve is shown in diagrammatic form in Fig. 35. It comprises, basically, a thin disc free to move inside a housing slightly larger in diameter and width than the disc. A large diameter:thickness ratio is normally chosen to eliminate cocking effects and minimise the mass of the disc for a given area.

In action the disc is free to move backwards and forwards through a small distance d, under the action of pressures acting on its faces. At extreme movement in either direction the disc forms a seal, whilst at intermediate positions there is flow around the periphery of the disc. The extreme positions represent the steady-state positions and thus the switching position is dependent on the relative values of P1 and P3. Ideally, at least leakage only occurs during switching transients.

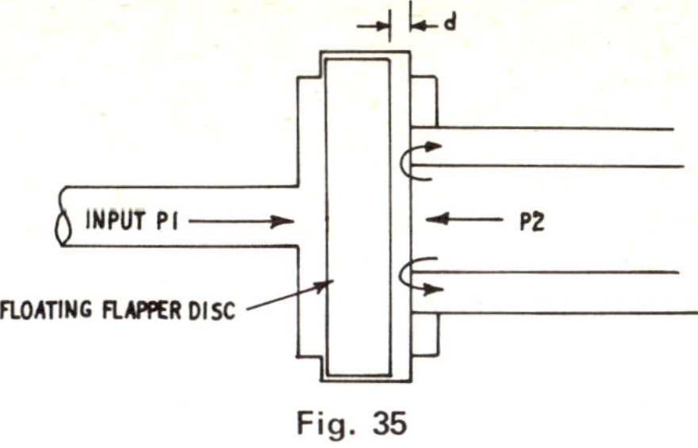

Fig. 35

Numerical Control Systems

Fluidic systems offer immense possibilities for the application of pneumatic control techniques, a particular example being the introduction of numerical control using fluidic components. In such cases control is established by low pressure pneumatics in conjunction with a block paper tape-reader, decoder, cycle timer and amplifier logic elements. These are coupled through high gain pneumatic relays to conventional air piloted hydraulic and pneumatic control valves.

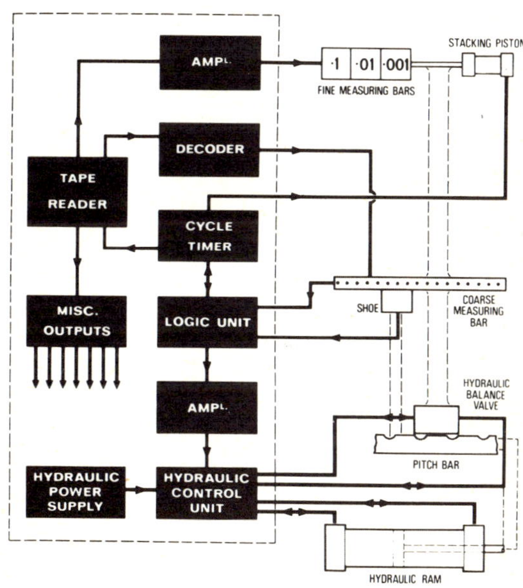

Fig. 36

A typical example is the Plessey Fluidic 220 system shown in Fig. 36. for precision control of table movement, table positioning being provided by a compound system of length bars and hydraulic servo. The programme is punched onto standard 8-hole paper tape. On command, a fluidic cycle timer causes the paper tape to advance one information block and the low pressure signals from the tape reader are amplified and used to set up the fine and coarse measuring systems. The cycle timer then stacks the fine measuring bars, unclamps the table slides and brings the air bearings and hydraulics into operation. The table then moves towards it's correct position at a rapid traverse, cutting out as then position is approached, when the servos are used to bring the table to the programmed position. The cycle timer then cuts the air bearings, locks the table and hydraulics, and signals the machine operation required.

Copying Devices

Copying devices for fitting to or incorporating in machine tools are an important field for hydraulic and electro-hydraulic servomechanisms. Types are available for fitting to most machine tools and they enable irregularly shaped parts to be machined economically, using either a template or a part which has been machined by conventional method to guide the tool.

The simplest copiers are those fitted to lathes, especially those producing multi diameter shafts. More elaborate copiers are fitted to diesinking machines. In all cases a stylus follows the contour to be copied and operates the servo valve. Simple lathe copiers are usually made as a unit with the valve incorporated, but on other machines the servo valve is separate.

DEVELOPMENT OF COPIERS

Non-electrical copiers have passed through several stages of development. Air jets have been used to sense the template with the object of avoiding physical contact with it and internal oil jets fitted to produce the necessary pressure differential to give movement. Many early copiers used what was in effect a hydraulic Wheatstone bridge. The stylus was attached to the valve spool which had the lands arranged in such a way as to give four resistances to the oil flow. The system would attain equilibrium when the ratios of P1 and P2 were equal to the inverse ratio of the effective areas of the two sides of the piston. It will be appreciated that this type of valve requires very careful manufacture

MODERN HYDRAULIC COPIERS

A single-axis hydraulic copier as fitted to a lathe (Fig.37) is normally mounted on the saddle which is traversed by the ordinary feed mechanism, either longitudinally or transversely, depending on whether the part resembles a shaft or a disc.

A lightly spring-loaded stylus maintains contact with the template and adjusts the position of the valve so that the cylinder and valve assembly move backwards or forwards on a slide – the piston is fixed in position.

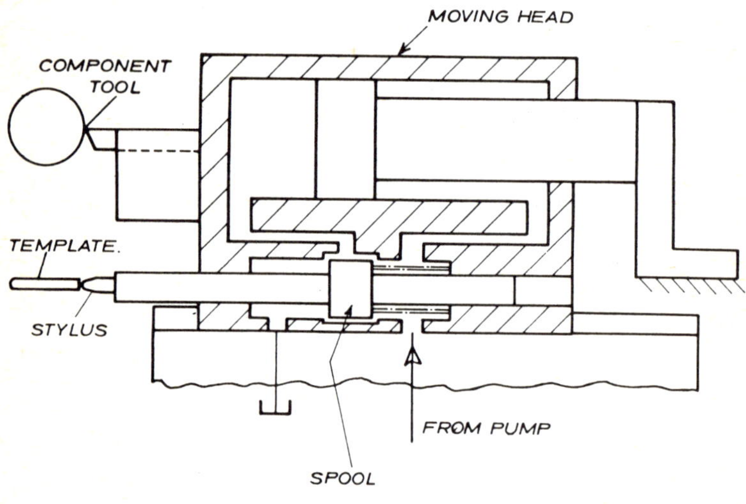

Fig.37:

Hydraulic copier head. The spool land acts as a hydraulic potentiometer, varying the pressure in the head end of the cylinder as required. When the head is stationary the head pressure is exactly one half the pump pressure.

The areas of the head and rod ends of the cylinder are in the ratio of 2 : 1. Oil at constant pressure is supplied from a separate power pack and flows continuously past the lands on the spool , so that when in equilibrium the pressure in the head end is half that in the rod end.

As the stylus moves along the template, changes in form cause it to move in or out a very small amount. This upsets the pressure balance across the spool lands and lowers or raises the pressure in the head end, causing the cylinder assembly to move until equilibrium is restored.

The action is analagous to an electrical potentiometer circuit,, where there are two hydraulic resistances R1 and R2 (Fig.38) in series, which are equal to each other in the equilibrium position. The pressure P2 (electrical or hydraulic) at the junction of the resistances is then exactly half the difference between the incoming and outgoing pressures, the latter assumed to be zero. As the effective areas of the two sides of the piston are in the ratio of 2 : 1 they must be in balance also. Any movement of the stylus causes one resistance to increase and the other to decrease and pressure P2 also varies causing the slide to move until equilibrium is restored.

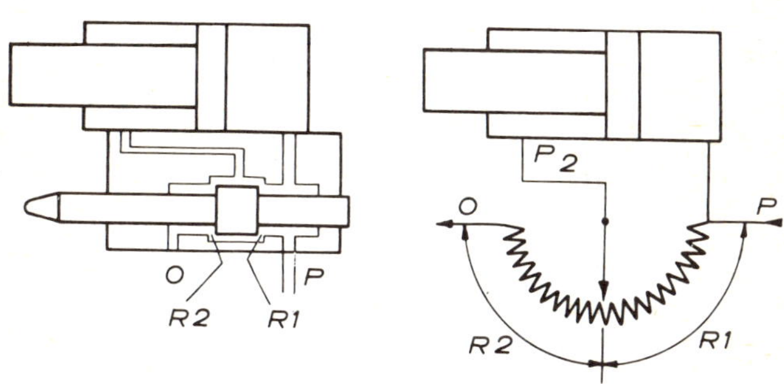

Fig. 38

Considering this as a not untypical servomechanism; firstly it is 'error operated' and no correcting action can take place unless the stylus moves slightly relative to the tool. One advantage of the spool breaking down the pressure in the way described is that great accuracy can be obtained without corresponding accuracy being needed in the construction of the valve. A commercial copier is sensitive to about 0.0002in.

The speed with which the copier head moves in or out to follow the contour of the template depends on the rate at which the fluid enters or leaves the cylinder and this in turn depends on the spool land position. It is obvious that the greater the slope of the template the greater the lag of the tool. For most commercial work the error is not likely to be important, but if it is, it can be allowed for by modifying the template.

Specifically, the slide velocity will vary with the stylus displacement and error. Provided the characteristics of the particular copier are known, then the slope of the portion of the template affected can be altered accordingly. Alternatively. an integral equalising network can be introduced in the system to eliminate any velocity error, so that the system can maintain a constant cylinder velocity

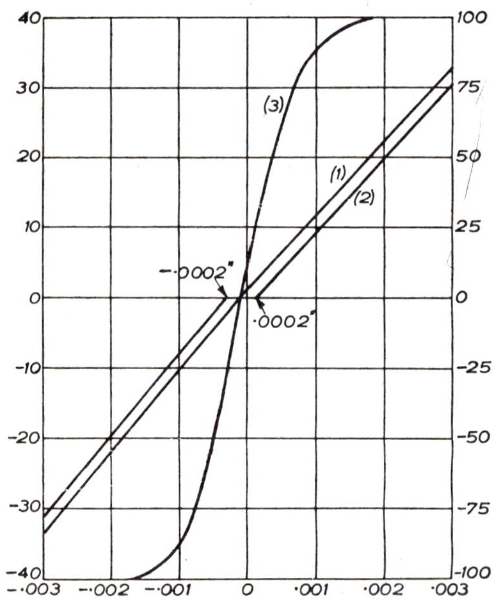

Fig.39

Graph showing how copying error varies with copier slide velocity.1. Velocity when unloaded; 2. Velocity when working; 3. Stalled force generated by hydraulic rams. (%).

with no error signal at the pick-off (i.e. no adjustment of the template slopes).

The steady state accuracy of this type of copier is of the order of .002 in and this gives some idea of the precision with which hydraulic valves can be made and their sensitivity to a slight unbalance. Like all servo-mechanisms it is 'error' operated and the reasons for this are probably more easily understood from this type of device than any other.

It is obvious that there can be no movement of the copier slide unless the stylus has to follow a change in contour. Whilst the movement is taking place, the valve must be displaced – this displacement is the amount of error. For a slow change of contour the displacement is small, but for a rapid change it must be correspondingly greater to pass the oil needed to give the necessary velocity. For a very rapidly changing contour a displacement of .002 in may be necessary. This can be clearly seen from Fig.39.

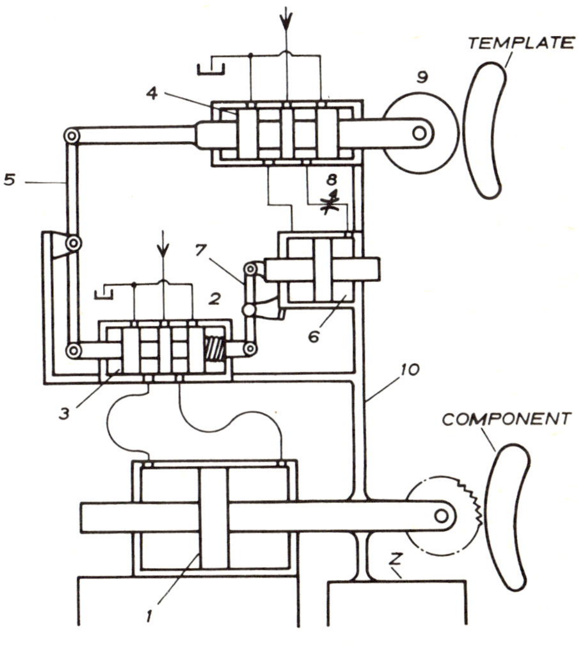

Fig.40

Hydraulic copier with integral correction device to allow for rapid changes in contour.

Although unusual on normal copiers, a specialised operation such as machining the profile of turbine blades may involve extremely rapid movements. Attempts have been made to overcome the resultant inaccuracy by using a 'reset', or integral, device which senses the movement of the spool and tends to compensate for it.

One method which has been proposed for doing this is shown diagrammatically in Fig.40, and this mechanism does illustrate quite well the principle involved. A milling cutter is being used for producing an aerofoil section by copying from a master which rotates at the same speed. The primary control valve 3 is operated through the lever 5 by the roller follower 9. The valve 4 is a secondary valve which is responsible for adjusting the primary valve 3 so that it opens more than it otherwise would when the follower is on a steep part of the slope. By doing this the error, which normally occurs when the primary valve has to be moved appreciably from its neutral position to produce the necessary velocity of the slide, is almost eliminated.

Unlike the previous copier cylinder, the areas of both sides of the piston are the same and the valve spool is of a slightly different design to allow for this. The most important difference, however, is that the valve is mounted in such a way that the body can be moved axially by lever 7.

The method of operation is as follows: A movement of the follower 9 causes the spool 4 to move and also the spool of valve 3, so admitting fluid to the appropriate side of cylinder 1 and exhausting it from the other side. Spool 4 also admits oil to one side of cylinder 6 in proportion to the movement of 9. This causes the body of valve 3 to be moved in the opposite direction to the original movement of the spool, so increasing the valve opening and speeding up the movement of piston 1. The speed of piston 6 is regulated by valve 8. As the copying error is mainly due to the acceleration rather than the velocity of the copier, this type of correction is known as integral.

Electro-hydraulic copiers have certain advantages for three-dimensional work such as die sinking. On these machines the tracer head must often be separated from the main hydraulics and should be capable of contouring over a wide range of angles. Provision is often needed for slow down and speed up when following steep contours. All this can be done better by an electric pick-up. In the Versa-trace system a 400 c/s circuit detects the errors of the tracing head and feeds them to an electro-hydraulic servo valve mounted on a hydraulic motor. The motor drives the table servos.

An error which is liable to affect all machine tool controls is that due to the phenomenon known as 'stick-slip' which is most apparent when slides are moving slowly. It is caused by the difference between static and dynamic friction. A force is built up to overcome the static friction which results in a slight compression of the various parts. When the slide starts to move, this force is released and the slide advances a small amount and stops until the force again builds up. Stick-slip can be overcome by hydrostatically loaded slides, where oil is forced through fine orifices into pockets cut in the slide-ways. The action is very similar to the jet device described previously, the moving slide forming the 'flapper'. Other methods include various designs of roller and ball slides.

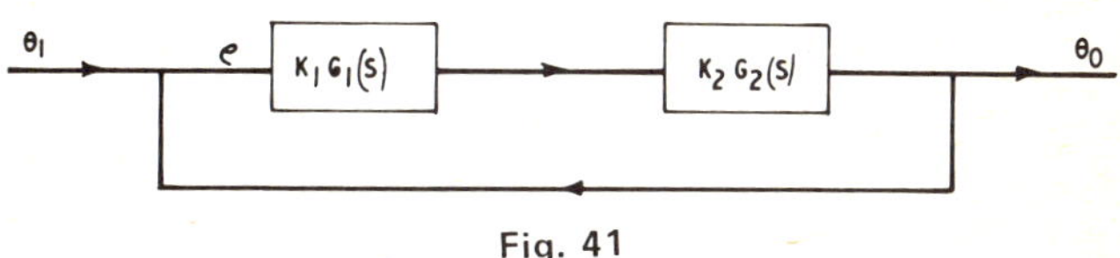

Fig. 41

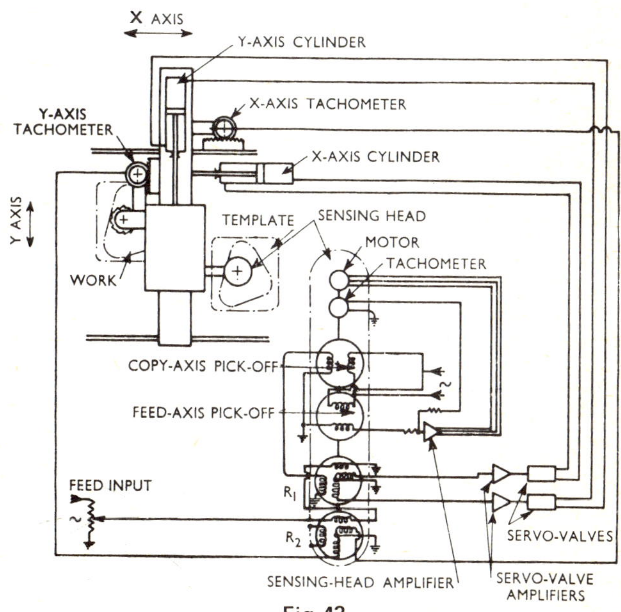

Fig. 42

(By courtesy of the Institution of Mechanical Engineers)

In the block diagram of a basic single-axis copier (Fig.41) block K_1G_1s represents the transfer function of the linear pick-off, amplifier and servo valve. The second block K_2G_2s represents the integration resulting from the conversion of flow from the valve to jack displacement. The minimum velocity error is governed by the loop gain and thus this must be set (at the appropriate frequency) for the required stability The addition of a further network providing an integral type compensation can greatly increase the stability, or enable a higher gain to be employed with negligible effect on the initial system stability.

A basic two-dimensional copying system (Fig.42) incorporates two controlled motions at right angles. The sensing head also has two orthogonal axes, one of which (the copying axis) is kept substantially normal to the template surface at the point of contact. The second is the feed axis. Feed motion is necessary to derive copying action, but at the same time a feed-axis pick-up detects any departure from a normal position, driving the sensing head in the presence of an error signal to maintain feed and copying axes at right angles.

The output from the pick-off in the copying axis is fed to one winding of a synchro-resolver lying in the same axis as the copying axis. The stator of the synchro-resolver has two windings, aligned with the two output motions respectively. The outputs from these two stator windings are then used to control their respective output motions via amplifiers and servo valves in the usual manner. As a result, deflection of the pick-off in the copying axis produces a movement of the tool in the same axis.

Feed speeds are read by separate X-axis and Y-axis tacho-generators mounted on each of the output motions, and feed error voltage determined as the difference between the resolved speed voltage derived from a second synchro-resolver, and the demand feed speed set by a potentiometer. This error signal is fed to the second winding on the first synchro-generator to provide synchronised feed motion. In the block diagram of this system (Fig.43) the open loop transfer function of the complete system is

$$\frac{\Theta_o}{\epsilon} = K_X G_X (s) \cos^2 \phi + K_Y G_Y (s) \sin^2 \phi$$

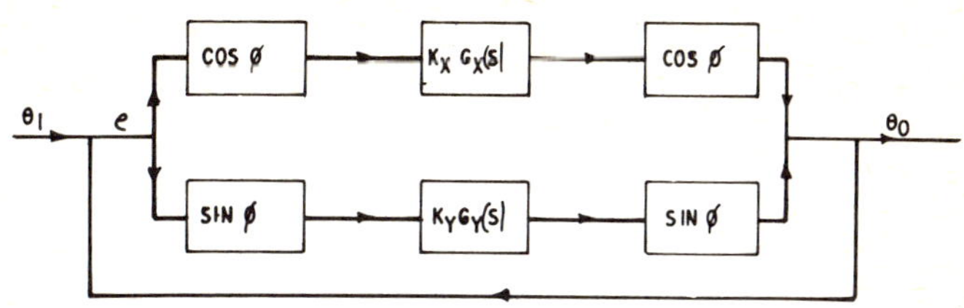

Fig. 43

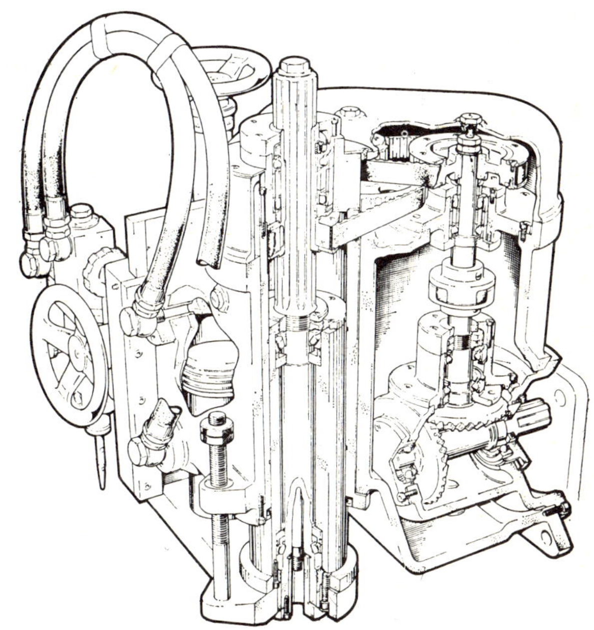

Fig. 44

Vertical hydraulic copying attachment for fitting to horizontal milling machine.

Copying on Milling Machines

Where copying does not provide full capacity for a single machine attachments may be used for fitting to existing machines, which can thus be used for ordinary work when there is no copying to do. Such an attachment is shown (in section) fitted to a horizontal milling machine (Fig.44) incorporating both the cutter spindle and copying mechanism.

The mechanism incorporates a pick-off stylus and ram, similar to that of a single axis-copier. A 3 : 1 speed increase gives sufficient speed for the small cutters. The stylus loading is only 6 ounces and this can be further reduced by a relay valve to 2½ ounces, if necessary, so enabling copying to be made from plaster or other non-metallic models. Provision can also be made for traversing the machine table automatically and applying a cross-feed at each reversal, so making operation automatic. The main table drive can be hydraulic cylinder, hydraulic motor or electric motor, whichever is most convenient. On milling machines designed especially for copying it is becoming common, where a cylinder is inconvenient, to move the table with a recirculating ball screw and nut mechanism driven by a hydraulic motor. This form of irreversible drive is very smooth and eliminates all backlash.

Tracing Heads

On machine tools other than lathes it is possible to have a separate tracing head connected by flexible hoses. These valves can be either three or four-way, depending on the system used.

For end milling on a vertical spindle machine the table may be traversed backwards and forwards, copying a three-dimensional model by raising and lowering the table by a vertical cylinder. The tracing head would then need to have up and down movement only and would be similar in action to that used on a lathe.

A tracing head controlling a single ram, can be made sensitive to displacement of the stylus in any direction, e.g. utilising ball connections give a vertical movement to the valve spool if the system is displaced sideways in any direction.

Another arrangement is needed to operate X, Y and Z axes independently. This involves a machine with cylinders which move the table in both horizontal directions and vertical. For some applications, however, the Z movement is omitted. A very small stylus movement is needed to give full speed control. In one system the stylus is held against the template by hand and the operator can also use the stylus to control the traverse independently of the template – as when passing over gaps.

The most searching test for a copying mechanism is when machining a slow taper. The slide must move forward gradually without 'stick-slip' whilst the valve is very slightly offset from the neutral position.

It is possible to use this mechanism to demonstrate a phenomenon which occurs with servomechanisms operating at high frequency – although rather unlikely to occur on a copier. Assume the template is in the form of a sine curve and the copier can be moved along it at various speeds; at fairly low speeds the tool follows the template faithfully (except for the small velocity errors mentioned previously). As the traversing speed is increased, not only does the tool movement tend to become less than template throw, but the peaks become later and later (Fig.45)

The reason for this is the time factor. At slow speeds there is sufficient time for the oil to flow through the servo valve to give the full movement to the tool. As the speed increases there is no longer

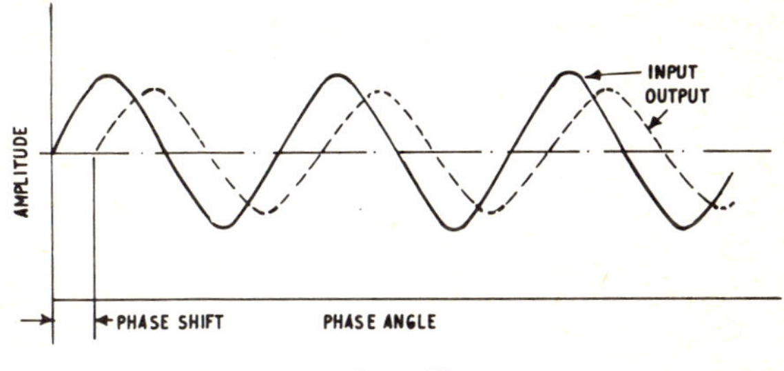

Fig. 45

time for sufficient oil to pass through the valve before it begins to close again and the tool does not travel the full distance. The interference with the normal flow also accounts for the fact that the tool will be sluggish in its response and its maximum movement will be somewhat later, relative to the template, than at slow speeds.

The rate of the square of the actual and possible movements is called the amplitude ratio and is a measure of the power the valve is capable of passing at that particular frequency. The lagging behind of the actual movement is called phase lag and is usually expressed in degrees. A phase lag of 90° means that the template is at its neutral point whilst the tool is at its maximum, and 180° that the tool is at a plus peak whilst the template is at a minus peak.

This phenomenum of phase shift, and variation in amplitude ratio, is typical of hydraulic servo valve, amplifier and pick-off system and are basic system stability characteristics.

The tendency to develop very complicated electro-hydraulic systems has been discouraged by the success of numerically controlled machines, especially as the latter do not need physical models to copy, which in themselves impose limitations on tracing methods. There are, however, special applications where it is worth while even making special machines with electro-hydraulic control.

Electro-Hydraulic Servo Valves

The electro-hydraulic servo valve is one of the most versatile power amplifiers known. Currents of a few milliamps can be used to control almost unlimited power and they can be responsive to input frequencies of 200 cycles per second or more. Technically they are high gain amplifiers and require a suitable electrical circuit to operate them.

The majority of electro-hydraulic servo valves have primary and secondary stages, the primary stage being operated by a torque or force motor. There are a number of different types of valve available, most of them originally developed for aerospace and automatic guiding systems but modified through the experience gained to suit commercial applications.

Electro-hydraulic servo valves are essentially high gain amplifiers which enable a very small electric signal to control unlimited power with a speed of response which seems to be only limited by the machine or mechanism being driven. They have even been made to reproduce 'musical' notes.

Commands are invariably electrical signals which may emanate from two different types of source. In one, as in a copier or speed control, 'error' signals are arranged to give a voltage output in proportion to the amount and direction of error. The other sets the pattern the valve has to follow in the electrical circuit itself, which may embody oscillating and timing circuits.

Servo valves are relatively inefficient devices. Assuming they are fed from a source of constant pressure, which means either that the pump discharges continuously through a relief valve or is of the pressure controlled type, there is a pressure drop through the valve of one-third the inlet pressure. In other words one-third of the power is lost in the valve itself. The reason for this is not readily apparent. As far as flow is concerned, a valve is merely a variable restrictor or resistance and it has been proved experimentally that with viscous flow, the quantity of fluid which is passed varies as the square of the pressure drop across the restriction. (Viscous flow is the normal type of flow, with the fluid in a state of turbulence).

Power varies as the flow and the outlet pressure; i.e. power varies as (pressure drop)^2 and (inlet pressure − pressure drop) or:

$$P \alpha \dot{\rho} (pi - \Delta\rho)$$

If a graph is drawn (Fig.46) with P as the vertical ordinate and $\Delta\rho$ (pi − $\Delta\rho$) is the horizontal ordinate, the relationship of power to pressure drop becomes immediately apparent.

This means that if a valve is used to its maximum power, i.e. a supply pressure of 3000 psi, there would be available 2000 psi at the actuator, with a flow depending on the size of the valve.

Fortunately, efficiency is not a major consideration for many of the purposes for which electro-hydraulic valves are most useful. Efficiency for a given flow or power demand can, of course, be improved by a larger valve, but this will involve the penalty of less precise control.

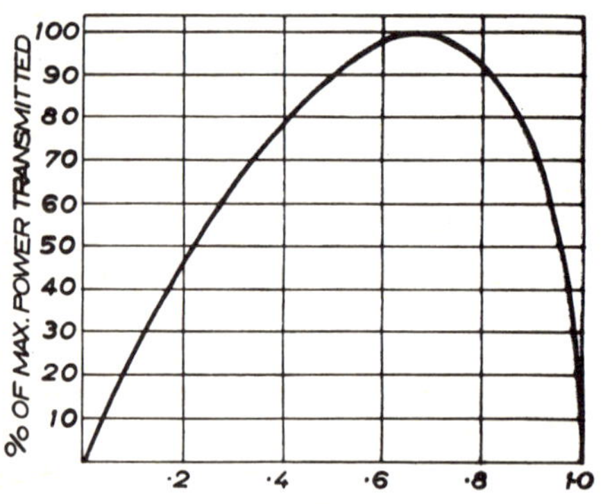

Fig. 46 :

Graph showing relationship of power and pressure drop through a valve.

ELEMENTS OF SERVO VALVES

The three pricipal parts of servo valves are:

1. The Torque or Force motor.

This is an electro-magnetic device having an armature which is actuated by two opposing coils. These are energised by a low voltage variable direct current having a maximum value of a few milliamps, which changes in polarity to reverse the direction of flow through the main valve. The force depends on the difference in the currents in the two coils.

The difference between a torque and force motor is that the armature is pivoted in the former whereas it moves longitudinally in the force motor.

Comparatively heavy permanent magnets give a high magnetic intensity. Because the magnets attract small ferrous particles it is usually thought preferable to separate them from the working fluid.

2. First Stage

Undoubtedly the most critical part of the valve, and one which most varies between different designs, is the first stage, as it must be sufficiently sensitive to match the torque motor and yet stable under practical conditions. There are two principal types of first stage construction.

(a) Port Type (e.g. see Fig. 47.)

The moving element may be either of the cylindrical (spool) or flat plate type and usually acts as a four-way directional control valve. The lap of these valves is very critical and requires accuracies of the order of .001in to obtain the desired characteristics.

Friction tends to be relatively high with a spool-type valve, requiring operating power levels of the order of 1 to 10 watts, which may be met by drive amplifiers applied to the torque motor. The flat plate type is possibly superior in this respect, although requiring extremely accurate lapping.

Internal leakage in the first stage of spool and flat plate valves is usually negligible compared with the quiescent leakage of the second or main stage.

(b) Nozzle Types (e.g. see Fig. 48.)

Servo valves which embody a nozzle system are claimed to be more tolerant to contamination than those with spools but, on the other hand, there is the disadvantage that the flow in the null position is somewhat greater.

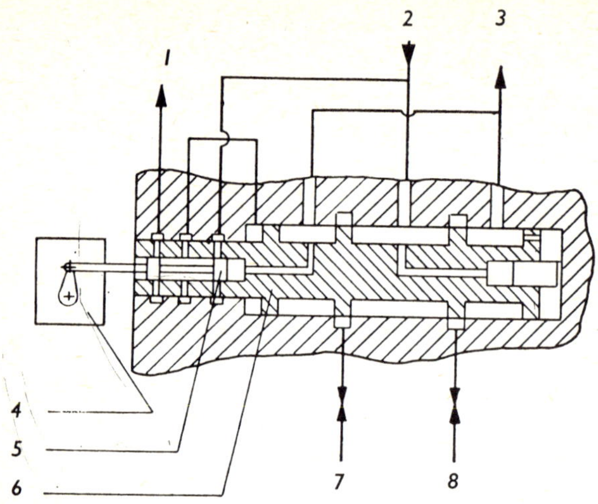

Fig. 47

Electro servo-valve with composite spool primary stage; 1. To tank; 2. From pump; 3. To tank; 4. Torque motor; 5. Pilot valve; 6. Main valve; 7. and 8. To actuator.

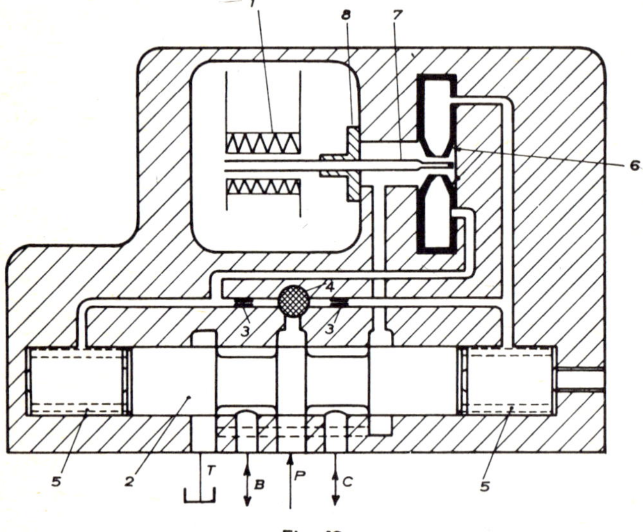

Fig. 48

Electro servo-valve with jet primary stage; 1. Torque motor; 2. Main spool; 3. Restriction; 4. Filter; 5. Centring springs; 6. Jets; 7. Flapper; 8. Flexible diaphragm.

Although designs differ considerably there are normally two opposed nozzles, each with its own restrictor jet. The latter is about half the diameter of the nozzle, say .020in diameter, and is therefore the most vulnerable point for choking up. If the size of this is compared with the annular gap of, say, a ⅜ in diameter spool of the same area

$$(.020)\,\frac{2}{4} = x\ .375\ x\ t$$

$$t = \frac{.0004}{4 \times .375} = .0003$$

the gap is of the order of .0003in and this is obviously a much more efficient straining device than the nozzle.

In a typical arrangement the two nozzles are in parallel configuration, directed at opposite ends of the pivoted torque motor armature or flapper. A differential current applied to the torque motor coils applies a torque to the flapper, balanced by the hydraulic reaction forces resulting in a discrete angular displacement of the flapper. In effect, flapper displacement provides variable orifice characteristics at each nozzle, or differential metering of the control flow. Control pressures for the second stage main valve are tapped off between the variable orifices and fixed orifices.

The principle advantage of the flapper nozzle type is that only very lower power inputs are required for satisfactory operation, e.g. of the order of 0.1 watt only.

A further nozzle type is the jet-pipe valve where a single jet only is employed, fed at system pressure. The pressure energy of the fluid is converted into kinetic energy at the nozzle exit, the flow then impinging on tapering passages directed to the main valve spool, being reconverted into pressure energy in these passages.

The principle advantage of jet-pipe valves is that they are less likely to become clogged or blocked than other types. They do, however, have a higher leakage and lower dynamic performance than other types.

3. Second Stage

The majority of second stage valves are of the four-way spool type and are little different in appearance from the ordinary directional control valves.

Some means must be provided for biasing the spool to the neutral position. This may be done by a spring at either end with means for adjustment by a direct mechanical linkage with the first stage, or electrically.

To give stability combined with sensitivity the spools are underlapped, contributing to the null leakage.

Different combinations of first- and second-stages yield valves with different overall characteristics. Thus as a general rule, a flapper-nozzle first stage yields a valve far more sensitive to changes in supply pressure than one where the first stage is of port type. Specific individual characteristics may also differ.

It is the usual practice for makers of electro-hydraulic servo valves to publish typical characteristics, usually in graphical form. One of the best guides to the performance of a valve is the graph showing the phase lag and amplitude ratio for various frequencies.

The reason why phase lag and reduction of amplitude occurs can be appreciated by considering the sample case of a copier mechanism following a template shaped as a sine curve. At low traverse speeds the tool will follow the template faithfully (except for velocity curves with which we are not concerned here) but as the traverse speed increases the hydraulic system, due to inertia and similar effects, will start to lag behind and as it does so the tool movement or amplitude will diminish. At the 90° phase lag shown in the graph the amplitude might have fallen to about one third of what it was originally. (Fig 49)

In a typical amplitude ratio/phase lag graph (Fig.50) the tests are made with a sinusidal current, the crest to crest displacement being taken as 360°. At 100 cycles per second the phase lag is 90° approximately, which means in effect that the valve spool is 1/400th of a second behind the input current.

The amplitude ratio is given in decibels, or rather –dB, indicating that the ratio is less than unity. This is another way of saying the scale is logarithmic and is based on the formula dB $= \frac{1}{20} \log_{10} R$. The equivalent ratios for various values of –dB are :– 0dB = -1dB = 0.89; –2dB = 0.794 –3dB = 0.708 –4dB = 0.63 and –10dB = 0.316.

Up to a certain frequency which may vary between 10 and 80 cycles per second, depending on the type of valve, the ratio is 1 : 1 and the flow through the valve at maximum current is the same as under static conditions.

The decibel, or logarithmic scale was originally introduced as it allowed values to be added (as in logarithms) instead of multiplying, though whether it is really necessary for commercial valves is debatable.

A better way of getting things into proportion is to realise that most electro-hydraulic servo valves connected (via suitable transformers etc) to a 50 cycle electricity supply would have little difficulty in responding very closely to the voltage variations.

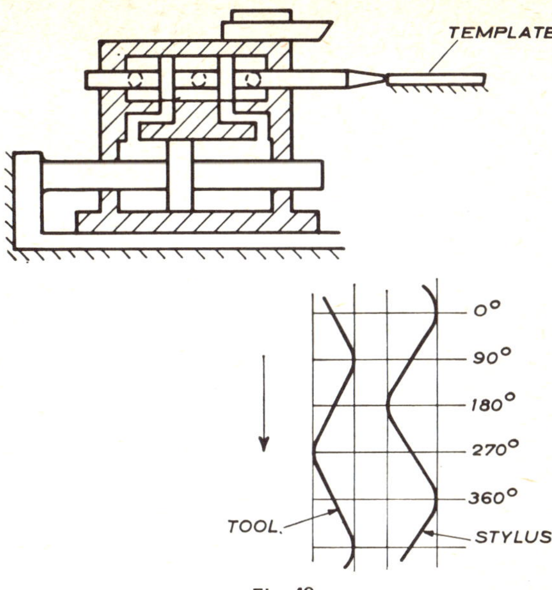

Fig. 49

Copier mechanism with graph below showing phase lag and reduction of amplitude for a very rapid traverse over a sinusoidal template.

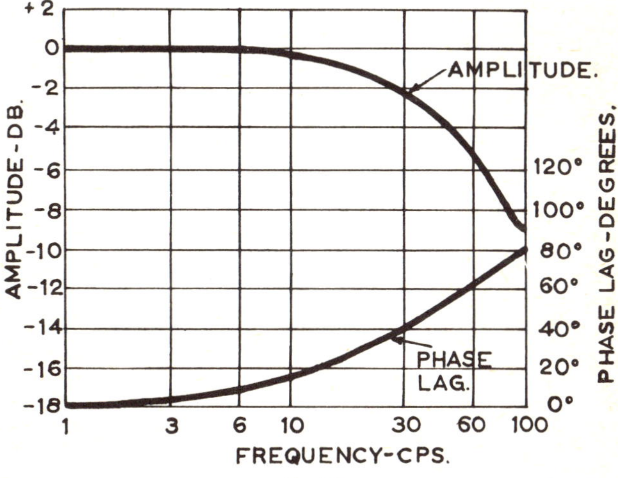

Fig. 50

Typical curve showing phase lag with frequency for an electro-hydraulic valve.

A user will, naturally, be concerned with the actual flow through the valve and this is usually given for a pressure drop of 1000 psi with the valve fully open. The flow for two smaller values of openings varies in proportion to the net electrical current applied. The figure of 1000 psi is chosen because it has been a common practice to operate military systems at 3000 psi and the valve passes maximum *power* when one third of the pressure drop occurs in the valve.

The valve is usually fed from a source of constant pressure, in practice either a pump discharging continuously through a relief valve or a pressure controlled variable delivery pump with or without a relief valve. The power transmitted by the valve depends on the flow and pressure differential. It is assumed that the flow through the valve is viscous or turbulent so that actually the flow varies as (pressure drop)^2 and (inlet pressure – pressure drop) or power $p(p_{in} - p)$.

If a graph is drawn (Fig46) with % of maximum power as the vertical ordination and $p(p_{in} - p)$ as the horizontal one, the relationship of power to pressure drop becomes immediately apparent.

This means that if a valve is to pass maximum power at the actuator, then the pressure drop through the valve must be about one half of the pressure differential across the actuator, i.e. $(p_{in} - p) = \frac{1}{2}\, p$.

It is often assumed that the inlet pressure is 3000 psi and pressure drop through the valve is 1000 psi.

Another way of looking at this is to consider what actually happens when the valve controls a motor which is intended to run at a constant speed under varying load. To simplify matters it will be assumed that an attendant checks the speed with a tachometer and regulates the current to the electro-hydraulic servo valve through a rheostat so as to keep a steady speed which is also the maximum which the motor will attain with full load and that particular valve.

When there is no load the pressure drop through the motor will be negligible and that across the valve a maximum (P_{in}). The flow through the valve depends on its resistance to flow; in other words the energy lost where the fluid passes through the gap between spool land and valve body. The valve will therefore be at a minimum opening for the flow required to give maximum energy loss. As the load on the motor increases, so the valve will have to be opened further to compensate for the lesser pressure drop across it. The power delivered by the motor will increase correspondingly until peak power is reached. At this point any further load on the motor will increase the differential pressure across it and decrease that through the valve. As the valve flow depends on the pressure drop and it can, of course, only open so much, depending on the size of the valve, the fall in pressure across the valve can only result in one thing – a drop in flow. In other words the attendant will be faced by the same problem as that of a motor car driver who finds that on a hill the speed drops despite the accelerator pedal touching the floor boards.

If operated at lower pressures than 3000 psi the flow falls off and at 1000 psi, probably a more realistic commercial figure, the flow is about 60% of that at 3000 psi.

A popular valve such as that used on machine tools has a normal maximum flow of 20 cu. in/sec. at 1000 psi pressure drop or 12 cu. in/sec. at 300 psi pressure drop (1000 psi inlet). This would be capable of displacing a 2in diameter piston against a force of 1800 lbs at the rate of 22in per minute; equivalent to about one tenth of a horsepower.

The largest valves, on the other hand, have an effective maximum output of about 10 horsepower at 1000 psi inlet pressure (45 horsepower at 3000 psi) entailing a minimum available power at the pump of say 20 horsepower (80 horsepower at 3000 psi) of which at least 10 horsepower (60 horsepower at 3000 psi) is rejected.

On the smaller valves the power involved is so small that efficiency is of no practical interest, but for the largest valves the overall, as distinct from peak, power consumption and type of power supply can be of major importance.

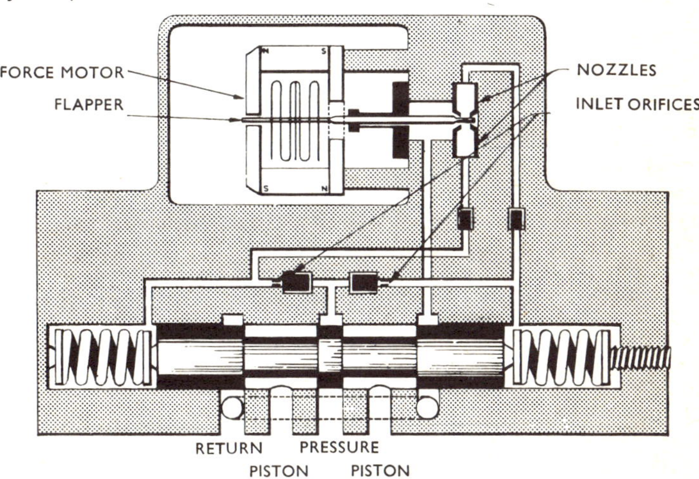

Fig.51 : Dowty Series 21 (Moog) servo valve.

MOOG VALVE

An electro-servo valve with nozzle or jet first stage is shown diagrammatically (Fig.51) The main spool is spring centred and in the mid-position the two outlets B and C are balanced. The first stage consists of a pair of nozzles fed through orifices from the main pressure

Fig. 51 cont'd.

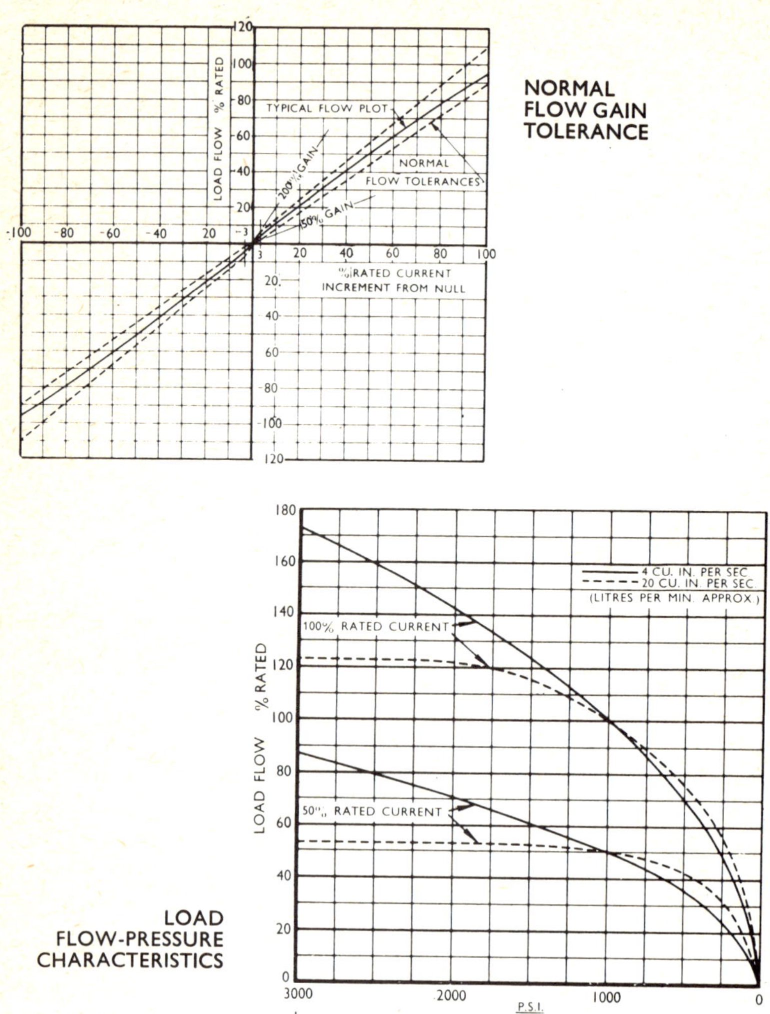

supply. A flapper between the nozzles is impinged on by both jets and when in the mid position gives equal jet back pressure, so maintaining the balance on the main spool. Any movement of the flapper resulting from a change in the amount flowing through the torque motor causes an unbalance in the jet pressures and a corresponding differential force on the main spool which causes it to move in proportion to the flapper control force against the resistance of the centralising springs.

With any nozzle and flapper type it is inevitable that there should be a constant flow of oil through the jets but the power consumed in providing an adequate oil flow is usually of little practical importance.

PEGASUS VALVE (Fig.52 and 52a)

Although this valve also employs nozzles for the first stage, the arrangement is entirely different to that described above.

The jets are incorporated in the second stage hollow spool whilst the pair of pivoted flappers move in unison with the force exerted by the motor. Any flapper movement causes the forces on the spool to become unbalanced and it slides until it is again centralised between the flappers in the new position.

For larger flows a second stage is available to which the first stage is attached. Both stages have AC pickups which provide a signal indicating the position of the valve. The second stage has a maximum capacity of 30 gpm for 1000 psi pressure drop.

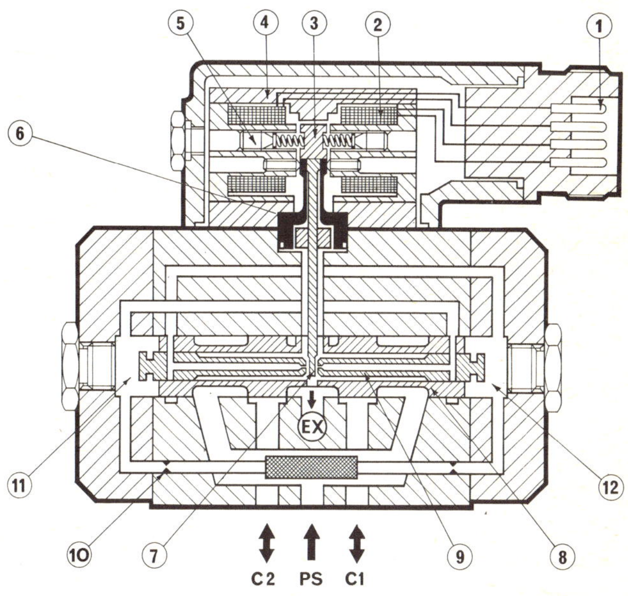

Fig.52 Pegasus Servo Valve. Key: 1. Amphenol receptacle; 2. Coils; 3. Armature; 4. Magnet; 5. Mechanical null adjustment; 6. Flexure tube; 7. Drive Arm; 8. Spool; 9. Variable nozzle; 10. Filter; 11. Boost chamber (PL); 12. Boost chamber (PR); PS – Pressure; Ex – Exhaust; (C1 and C2) Service Ports.

VICKERS SINGLE STAGE VALVE (Fig. 53)

This single stage valve primarily intended for controlling variable delivery has a sensitivity less than that of those described above. It consists of a rotary four-way valve having its spindle mounted on ball bearings. At the upper end is an arm having a coil moving between substantial permanent magnets. at the lower end is a helical spring which, when the valve is used for controlling a variable delivery pump, gives a feed back in proportion to the pump stroke. When controlling an actuator direct the spring is fixed. A similar spring mounted above it is used for zero setting or for manual override.

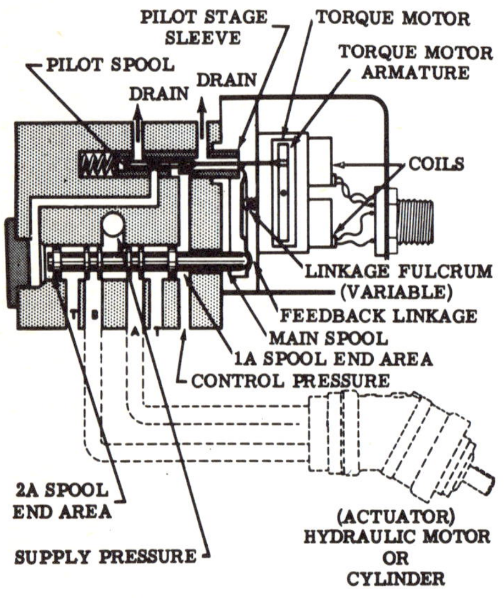

Fig. 53

FAIRY SINGLE STAGE VALVE (Fig. 54)

In marked contrast, this valve has a plate sliding between two flat faces. Great care is taken with the lapping of the faces to optical flat if necessary. The parts are pressure loaded through the plungers incorporated in the connections. The available output of the torque motor is sufficient to give a maximum flow of 1.86 gpm at 1000 psi pressure drop. The Bernouilli forces opposing the torque motor automatically limit the flow.

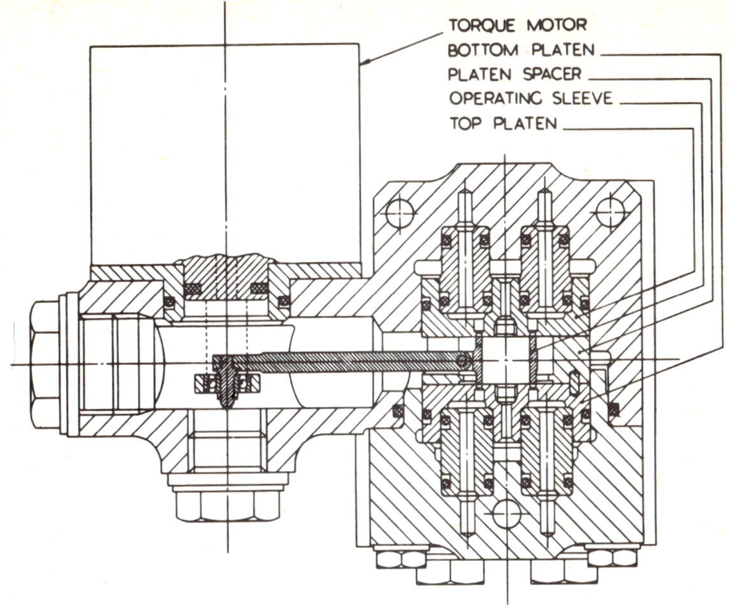

Fig.54

Fairey flat face servo valve.

Both single and two-stage valves are produced. The commercial type of single-stage valve is designed to operate with system pressures up to 4000 psi, though it is possible to operate at 5000 psi for short periods, if necessary. The flow which can be handled in a single stage is limited to 1.8 gpm irrespective of the operating pressure.

Two-stage valves have no Bernoulli force limitations, the flow depending on the supply pressure and the valve displacement from the centre at any particular instant. The flow characteristics of the two-stage valve depends almost entirely on the main hydraulic valve in the second stage. These second stage valves are based on existing Fairey type 902 and type 1137 hydraulic servo valves, with nominal flow ratings of 24 gpm and 61 gpm, respectively.

Keelavite (Fig.55)

This valve is made in three sizes offering flow rates from 10 gpm to 120 gpm for a pressure drop of 1000 psi across the valves.

This two-stage valve is of the spool-spool type, the pilot valve being of three way construction. A limited angle torque motor drives a crank arm against which the two single orifice spools are centred by means of springs. A rotary AC pick-off is also driven by the torque motor, as shown in Fig.55.

Two balanced windings are fitted to the torque motor and when the current fed into these coils is equal, the valve is held in the centre position by means of the springs. If a differential current change takes place, the spools are deflected to either left or right, the deflection and resultant flow being proportional to the differential current.

SERVO CONTROLLED PUMPS

When the continuous power consumption exceeds about 5 horsepower, serious consideration must be given to a servo controlled variable displacement pump. The power output is then only restricted by the available pumps, but the sensitivity is reduced by the inertia of the pump displacement control mechanism. The usual method is to use a cylinder fed from the servo valve, using oil pressure from a pilot pump.

PUMP CONTROLS

In the system adopted by Vickers Ltd, the servo valve (Fig.4) is mounted coaxially with the pump control shaft. The torque motor turns a rotary valve through a helical spring whilst a similar helical spring is turned by the control shatf as it rotates into a new position. When the two balance the pump delivery agrees with the command signal. It is claimed that the time from zero to full stroke is 0.05 seconds and the flow is accurate within 0.1 per cent of full flow.

Other systems may employ a signal from the actuator giving speed or position to operate the servo valve.

The flow from a variable delivery pump is far less affected by variations of load or oil viscosity than when the flow is regulated by a valve, and for most practical purposes it can be assumed that the flow is constant for any particular position of the control. Overload should, however, be avoided or provided for, as it may affect the speed of the driving motor. Such a drive has obvious attractions for a machine tool feed where it overcomes the oil temperature affects so often troublesome with flow control valves.

As shown in the diagram, the two ends of the main spool differ in area by a factor of 2:1, the smaller area being fed with the same constant supply pressure as that applied to the pilot valve. With the pilot valve in the centre position, a small leakage flow occurs from Pressure to Jack to Tank and as the pressure and exhaust spools have same equal underlap, the pressure in the jack port is $\frac{P_s}{2}$. Consequently the net force on the main piston is

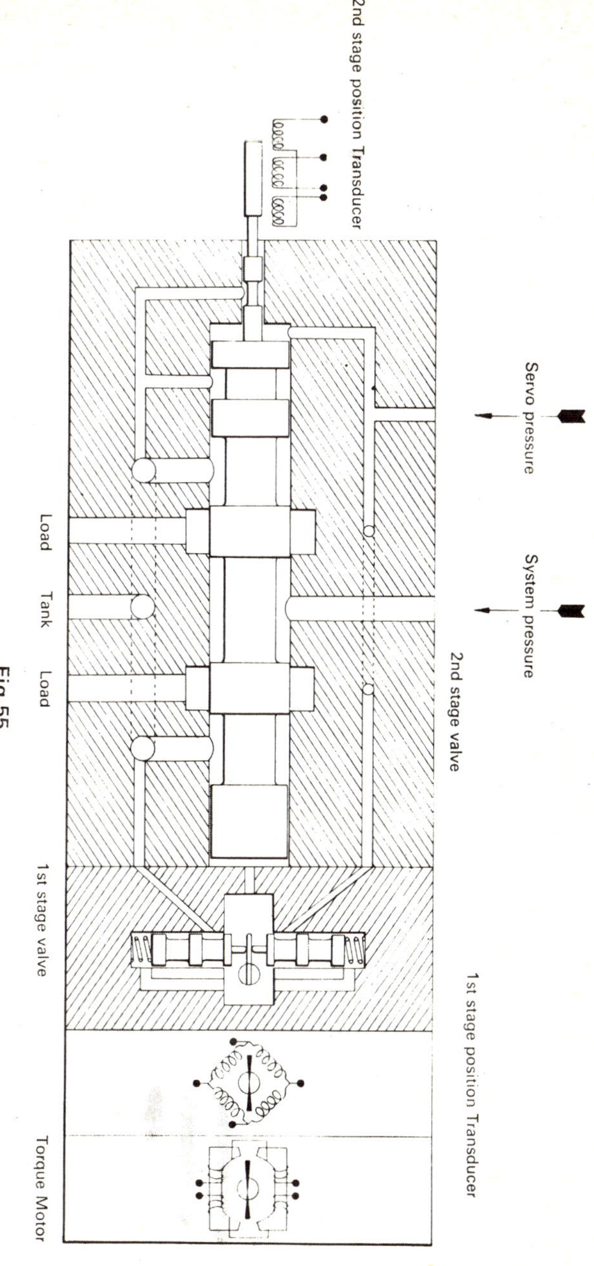

Fig.55
Keelavite two-stage valve.

$$Ps \times A - \frac{Ps}{2} \times 2A = \text{Zero}$$

If the pilot valve is deflected, the pressure in the jack port changes and the force balance is disturbed, with the result that the main piston moves at a speed proportional to pilot valve deflection, its direction being determined by the direction of pilot valve movement. It is desired, however, that the position of the main valve is proportional to the electrical input signal and an electrical position loop is therefore closed around the main valve.

ELECTRO-HYDRAULIC SERVO VALVE ELECTRICAL ACCESSORIES

A number of electrical accessories are necessary for the operation of these valves, which include –

Power packs which are suitable for connecting to the mains which produces both a direct current for operating the valve and a 400 cycles per second (H2) A.C. supply for inductive detectors and, when desired, to provide a 'dither' superimposed on a D.C. signal.

Linear pick-ups for detecting small differences of position – up to ∓0.025in and operating from 400 H2 A.C. By operating from high frequency A.C. sliding contacts are avoided and there is maximum sensitivity and response is approximately linear.

Rotary pick-ups for detecting small angular changes up to ∓ 30°. Operated from two H2 supply and having approximately linear response.

Synchros. A synchro is driven mechanically by each of two motions it is desired to synchronize. The two synchros are connected by three wires. Any discrepancy produces a signal which can be fed to the valve.

Tacho-generators. A D.C. generator which produces a current in proportion to the speed. (See Fig.56)

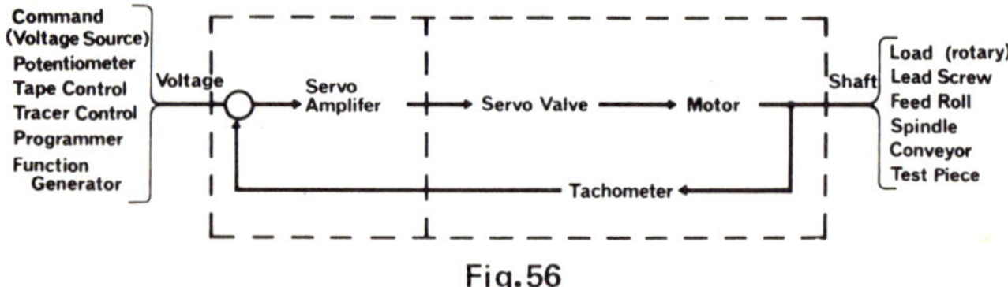

Fig.56

Telehoist servo drive with tachometer.

Potentiometers. One method of regulating a direct current which can either be fed into a servo-valve or used as standard against which the current generated by a tacho-generator can be compared and any discrepancy used to operate the servo-valve.

Amplifiers transform 400 cps A.C. signals to D.C. for feeding to the valve. D.C. signals may also require amplification.

Applications of Electro-Hydraulic Servo Valves

Electro-hydraulic servo valves were first developed for military purposes such as gun aiming and aircraft power control systems. They are also, in their commercial form eminently suitable for many civil purposes, where they offer entirely new concepts for both control and construction. It may often require a complete break with traditional methods to use the valves but the gains can be great. The following examples show some of the ways in which they are, or might be, used.

Electro-hydraulic servo valves are essentially modulating directional control valves, the magnitude and direct of flow being proportional to an applied small current which may vary 250 times per second. It is probably this property of what in practice amounts to instant response, which is the most difficult to appreciate, and yet which is likely to have the greatest potential.

Electro-hydraulic servo valves are principally employed in applications of the following types:—

1. To maintain a set position despite changes in load.
2. To maintain a chosen relative or actual speed despite changes in load.
3. To follow a set programme.
4. To follow signals sent out by instruments.
5. As a distance controlled variable flow regulating valve.

Normally the valves are controlled by several signals which may be A.C. or D.C. These are coordinated electrically and the result appears as a direct current which may be reasonably steady or fluctuate so that it is in fact alternating, and this is fed to the valve torque motor.

SHIPS' STABILIZERS

This well known contribution to the comfort of passengers and crews is an excellent example of an electro-hydraulic servo control responsing to a number of different signals. The servo valve itself is rather different from those described in the previous chapter and is intended to operate at a much lower pressure. This does, however, ensure the reliability essential for a mechanism which must operate continuously for weeks at a time.

Like most servo-mechanisms, stabilizers are error operated and cannot, therefore, eliminate roll entirely, nor unfortunately, are they effective if the ship is stationary.

It will be appreciated that if the wave on the port side of the ship is exerting a force which will cause it to roll to starboard, whilst the fins can be tilted to exert a righting couple. The effectiveness of the fins depends largely on the speed with which they can be turned from full tilt to full tilt in the other direction, and this time is about one tenth the rolling period of the ship.

The fins themselves may be retractable or non-retractable, to suit the type of ship. If retractable, they are moved in and out hydraulically when the ship is in harbour. The fins are tilted either by pairs of cylinders or by vane motors, the latter being a later development and has the advantage of taking up less space.

CONTROL SYSTEM

The control system is based on two gyroscopes, one sensing roll angle and the other roll velocity. The servo control gear was originally developed by Muirhead Ltd. for the Admiralty. On large ships, roll angle roll velocity, acceleration and natural list together with fin position feed back are taken into account. The acceleration, which is a derivative of velocity, is most important since it makes it possible to anticipate the roll and causes the fins to be adequately tilted before the roll has time to develop. The natural list control saves power as it enables the roll to be about the ship's natural list angle, rather than trying to keep it upright unnecessarily. On small craft it may be sufficient to use roll angle only as the control signal.

The practical effects are that a natural roll of about 30° out to out may be reduced to 3° out to out. The fin position feed back ensures that the fin movement is cancelled once the predetermined angle of tilt is reached.

ELECTRO-HYDRAULIC SERVO

The signals mentioned are coordinated by a transistorized circuit and operate a magslip which exerts a force of the order of 0.1oz. in and this in turn operates a two-stage self-contained hydraulic servo which has its own pump at 115 psi, a proportion of the output being passed through a reducing valve set at 40 psi for pilot valve and servo cylinder. Both stages have spool valves with resettable linkages. A dither mechanism for the first stage is formed by oscillating the cylinder. This in turn transmits a corresponding dither in the later stages of the system. This eliminates the effects of the friction and backlash.

The output is taken to a shaft which oscillates through a maximul angle of $\mp$ 11° with a maximum torque of 40 lb ft, i.e. a gain of 7680.

It is usual to operate the fin cylinders or vane motor through a variable delivery reversible pump (or pumps), the servo being connected to the pump control lever, on very small installations, or through a third amplification stage for large vessels. The system operates at about 300 psi and is supplied from a pump driven from the same motor as the fin tilting pump. To save power this pump may be of the pressure controlled variable delivery type. (see also Fig.57).

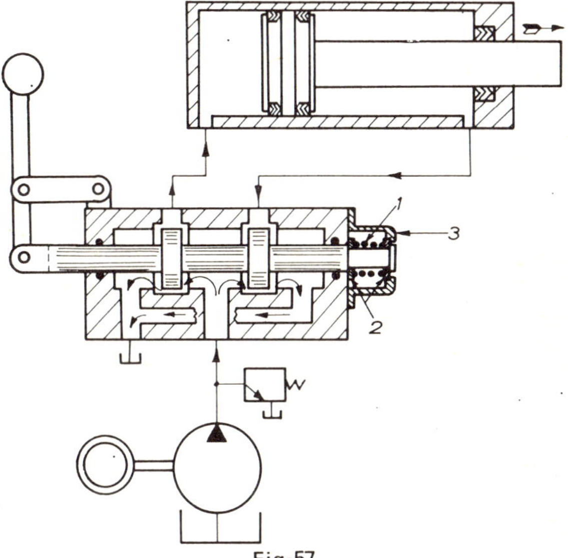

Fig-57

A three-position 'open centre' directional control valve 'unloads' the pump and allows it to discharge freely when the hand lever is released. 1. Centering spring. 2. Washers. 3. Spring cap.

A. Watertight box allocated to
F. Fin (retracted)
H. Hydraulic cylinder
X. Crossload
L. Lever

WAVE MAKING

Electro-hydraulic servo valves have been found just as invaluable in the control of the mechanism for producing waves artificially in ship testing tanks. The waves are produced by one (or more) wedge shaped plungers which are moved up and down to a predetermined programme to generate the height and pitch of waves required to test the particular mode.

In the DSIR Ship Hydrodynamics Laboratory the 1300 foot long tank has a single plunger 17 feet high, 48 feet wide and 7 feet wide at the top. It weighs about 20 tons, but being hollow, it is buoyant.

Electronic control gear determines the programme which is set for wave frequency and amplitude by dials on the control desk.

HYDRAULIC SYSTEM

The wedge is moved up and down by two pairs of double-acting, double-rodded cylinders, the latter feature ensuring symmetry in the control circuit. The natural frequency of the system had also to be taken into account as this governs the possible degree of control. The volume of active oil in the system was kept to a minimum and for high frequency both cylinders are connected in parallel to increase the natural frequency. For low frequency operation where cylinder stroke and velocity are at maximum one cylinder only is used.

Power for the cylinders is provided by four variable displacement reversible pumps, each delivering a maximum of 110 gpm at 1800 psi. One pump is connected directly to each cylinder for low frequency operation and two pumps to one cylinder for high frequency operation.

SERVO CIRCUIT

There are four Keelavite two-stage electro-hydraulic servo valves, one being mounted on each of the four pumps. The servo valves are connected to the displacement control pistons of the pumps.

The electrical system provides for the regulation of the servo valves to give the correct pump setting at all parts of the stroke of the plungers and also, the required length of stroke. The actual plunger position is continuously monitored and compared with the correct position as set by the dials and if any error exists the current to the valve is adjusted accordingly.

The actual circuit also includes a number of solenoid operated directional control valves which are essential to safety in event of an electrical failure and also for selecting the various modes of operation. The complete hydraulic current is shown in Fig 58.

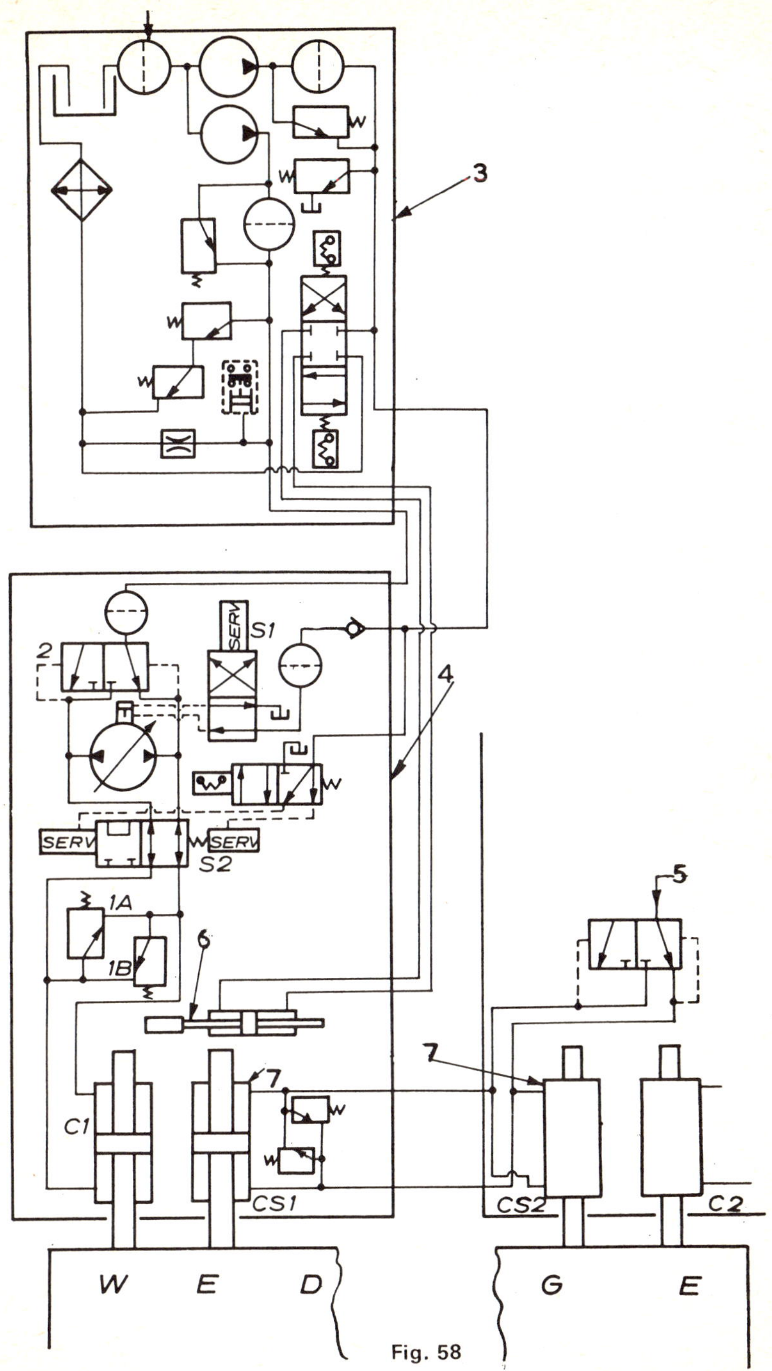

Fig. 58

RAPID TRAVERSE OF HEAVY MASSES

The problem of moving heavy masses over short distances is rather different from that involved in the previous case. In practical examples, such as the manipulation of forgings, it is usually necessary to move the mass a set distance and then stop whilst some other operation takes place.

Although it might be possible to achieve the desired result by a fixed displacement pump and suitable valving, there would be great danger of shock loads if the power necessary were applied suddenly by the opening of a conventional valve.

The availability of servo-controlled variable displacement reversible pumps of sufficient capacity makes it possible to provide the required sequence of acceleration, constant speed and deceleration with absence of shock and much more efficiently, whilst no large control valves are needed. One such application is for moving the 1 - ton magnet of a Proton accelerator The magnet is moved through a distance of 20 inches in a time which may be as short as 0.35 sec. and power is supplied by a pump with a nominal output of 250 horse power, although in this instance it is powered by a 120 horse power electric motor. The short time full pressure is required allows the additional power to be supplied by the motor inertia (fly-wheel effect). It will be noted that a boost pump feeds the low pressure side of the system and that a relief valve gives protection against excessive pressure on the high pressure side. The servo valve operates for a separate 2500 psi circuit with its own pump and accumulator.

The acceleration and deceleration of the mass depend entirely on the cylinder area and maximum pressure allowable, whilst the steady speed depends on the pump output. The power consumed is greatest when accelerating and falls off sharply when moving at a steady speed which it does for the major part of the traverse length. It will be seen that the predominant factor in determining the time is the capacity of the pumping system, and to halve the time means doubling the capacity.

In an installation of the type shown (Fig.59.) it is assumed that the mass weighs 3000 lbs, which is moved by a cylinder with an effective

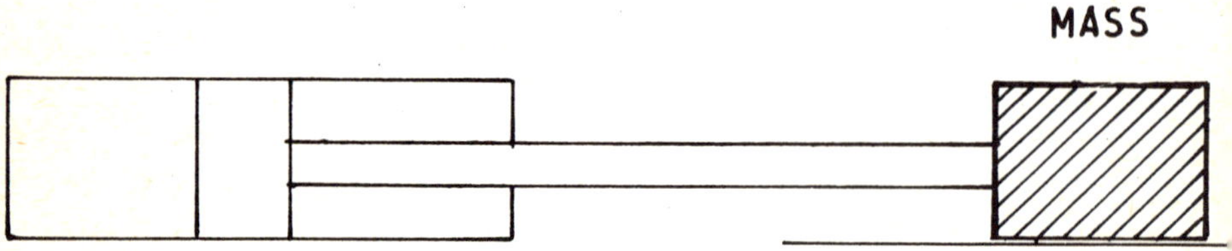

MASS

Fig. 59

area of 9.3 sq. in. The maximum pump delivery is 360 cu. in/sec. and the maximum pressure 3500 psi.

Acceleration is calculated from the maximum cylinder thrust –

$$3{,}500 \times 9.3 = 32500 \text{ lbf}$$

or

30000 lbf if friction is allowed for.

$$\text{Acceleration} = \frac{\text{force}}{\text{mass}} = \frac{30000 \times 32.2}{3000} = 322 \text{ ft/sec}^2$$

The terminal velocity is given by dividing the maximum pump output by the cylinder area –

$$\frac{360}{9.3} = 39 \text{in/sec or } 3.24 \text{ ft/sec.}$$

The time for acceleration is given by –

$$t = \frac{\text{final velocity}}{\text{acceleration}} = \frac{3.24}{322} = 0.01 \text{sec.}$$

Distance travelled during acceleration –

$$= \tfrac{1}{2} \times \text{acceleration} \times \text{time}^2 = \tfrac{1}{2} \times 322 \times (0.01)^2$$

$$= 0.016 \text{ feet.}$$

It is assumed that acceleration and deceleration are the same.

Distances travelled at 3.24 ft/sec = 1.67 – (2 × 0.0161) = 1.638 feet.

$$\text{Time taken} = \frac{1.638}{3.24} = 0.49 \text{ sec. Total time} = 0.510 \text{ sec.}$$

To achieve this without shock the pump delivery control must be moved from zero to maximum in 0.01 sec, kept at maximum for 0.49 sec and then returned to zero in 0.01 sec.

During the acceleration the speed of pump and motor is reduced by about 5% and this is partially restored by recuperation during deceleration.

To achieve this the electro-hydraulic servo valve is programmed to operate the pump control at the calculated rate at the beginning and end of the stroke.

The same system and a similar circuit are employed for operating hydraulic motors which are more convenient when the distance is greater or the position of travel may vary. The motor itself must have a positive drive – such as gearing – to ensure there is no slipping during acceleration or deceleration.

VIBRATION TESTING

The testing of components and assemblies by simulating the vibrations and loads to which they may be subjected in service can give valuable information on their likely behaviour and enables modifications to be made in the development stage. This is of course ,very essential on aircraft and it is being realised that it is equally applicable to vehicles and structures. Dynamic loads may have very different effects to static loads, both because of fatigue and the effect on joints etc.

The majority of vibration testing machines except the smallest, are now based on a hydraulically reciprocated table which is connected directly to one or more electro-hydraulic servo valve which are controlled either by an oscillator to give a definite frequency or through a programme controlled which is arranged to simulate as closely as possible the actual service conditions.

The Dowty vibrator(Fig. 60)has a double rodded piston which is fixed to the base and drilled to take the fluid passages. The cylinder is integral with the table and forms the moving member. The servo valve is mounted on the base of the machine and acts as a directional control valve. It is fed from a separate power pack and controlled from a separate console. The position of the table at any moment is dictated by an AC inductive pick-up.

Control signals to give the desired frequency and amplitude of vibration for the particular test and generated on the console in conjunction with the feedback from the AC pick-up, which amongst other things, ensures that the table oscillates about its mean position despite the weight of the test piece.

By connecting the servo valve directly to the cylinder and keeping the connecting passages as short as possible the full capabilities of the valve can be employed.

On an application of this type, acceleration is directly proportional to the differential pressure across the piston and velocity to the quantity of fluid. If the usual assumption is made that one third of the pressure drop is in the valve, then the centre of the stroke is governed by this. It may be rather more at the beginning of the stroke where the flow, and therefore the pressure drop, could be less.

For deceleration the problem is rather different as fluid must be pumped into the 'trailing' side of the piston to overcome the restriction in the valve and at the same time the exhausting fluid must be throttled. This involves a strong possibility that the pressure in the exhausting side of the cylinder is appreciably higher than the pump pressure and this must be allowed for in the choice of maximum pump pressure and the design of cylinder and servo valve.

The frequency and stroke are inter-related and high frequencies are only obtainable with a short stroke. A stroke of ∓ 2" might have a maximum frequency of 150 c/s. Vibration testers have been made with strokes of 7 feet.

Besides the position feedback already mentioned it is possible to have a force feedback from a load cell between the vibrator and specimen.

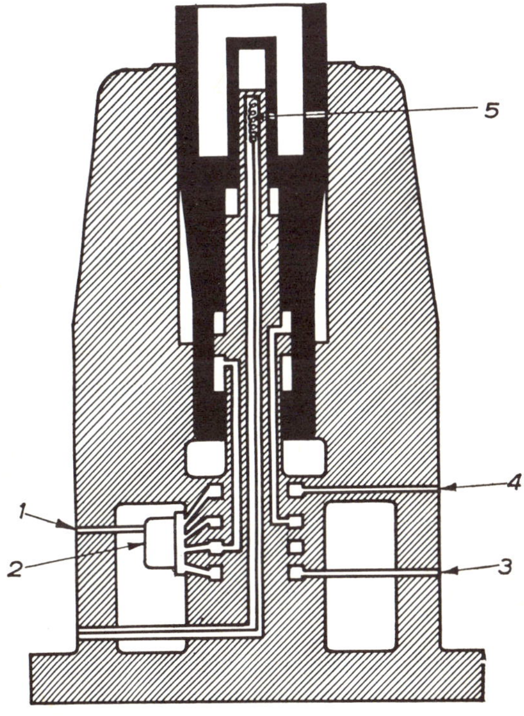

Fig.60

Dowty vibrator in section. 1. Input signal; 2. Servo-valve; 3. Oil pressure; 4. Exhaust; 5. Position feed-back transducer.

The force exerted depends on the area of the piston and the differential pressure across it, and as mentioned above, the pressure drop through the valve at the maximum velocity limits the available thrust.

SPEED CONTROL

All automatic speed controllers, from the Watt governor onwards, are servo-mechanisms, but such controllers are limited in their scope and are little used except for prime movers.

Electro-hydraulic servo valves make available several methods of speed control which are applicable to machines in general and which are suitable for both rotary motion involving hydraulic motors and linear motion by cylinders.

The simplest method of speed control is by using the servo valve as a distance controlled flow regulating valve. The opening of the valve is determined by setting a potentiometer which can be done manually or automatically if speed changes are required during a cycle. Such an arrangement would be subject to the usual fluctuations due to load and temperature changes of an uncompensated flow regulating valve. (See Fig. 61.)

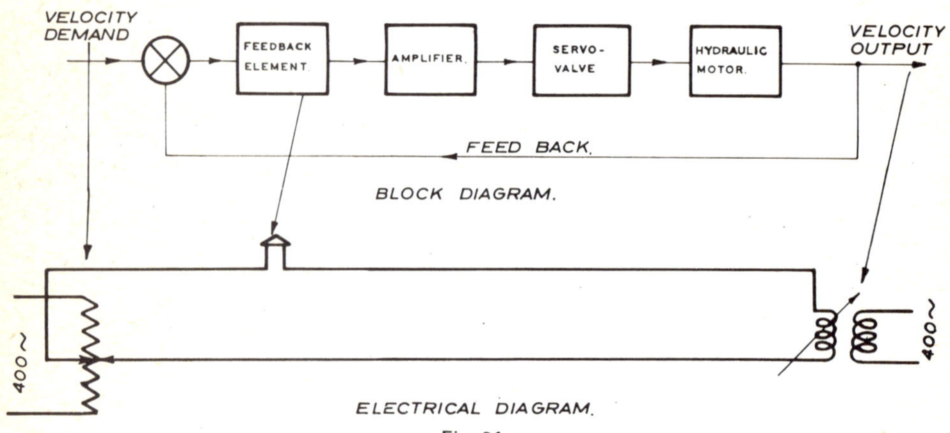

Fig.61

Block diagram of speed control with electrical diagram below to show how the voltage from the tachometer is subtracted from the voltage set on the potentiometer, the difference being the 'Feedback' element.

A simple pressure reducing valve connected across the servo valve will eliminate fluctuations due to variations of load and also reduce the pressure drop across it. The reduction of pressure has the advantage that the valve is wider open for a given flow and is therefore not so critical in adjustment. This scheme is probably the most economical of any electro-hydraulic servo valve application and is satisfactory where speed variations of 1% or so are permissible.

A possible method of varying the speed of a piston during the forward stroke and returning it at fast speed, is by using a rotary type servo valve. The speed is predetermined by a suitably shaped plate cam against which an arm attached to a rotary potentiometer is pressed by a spring. As the potentiometer moves with the piston along the cam its value is altered to conform with the speed required and at the end of the stroke it flies over to the other track, and the direction automatically reverses.

A more elaborate electrical arrangement is involved when a feedback is taken from the speed as measured by a tachogenerator (Fig. 62) or from the system pressure or load by means of a transducer and this

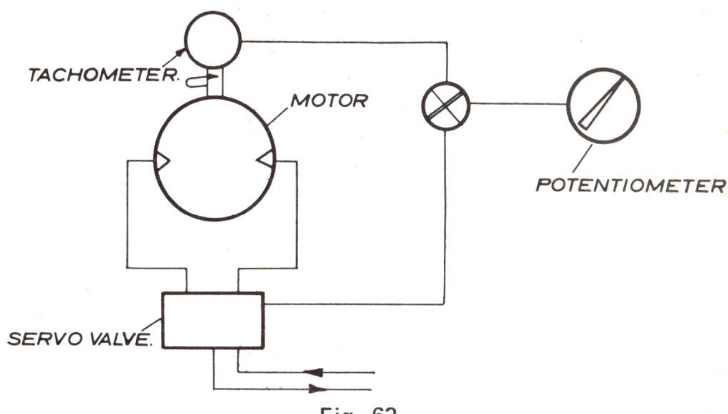

Fig. 62
Controlling hydraulic motor speed with servo-valve by signals from tachometer or tacho-generator.

is used to modify the potentiometer current through an amplifier. Various other methods and applications of speed controls are illustrated in Figs.63 - 65.

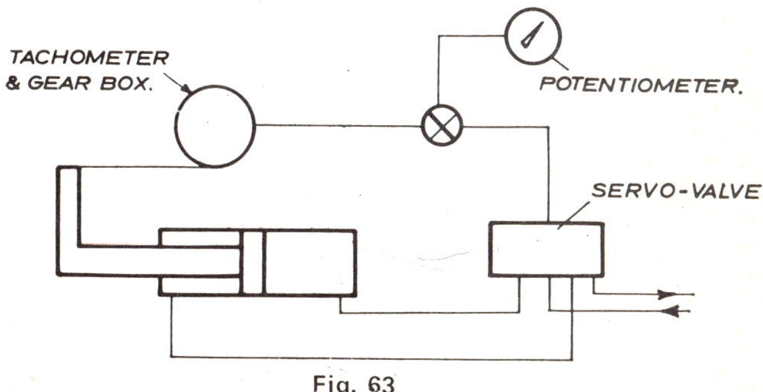

Fig. 63
Ram with servo-valve speed control.

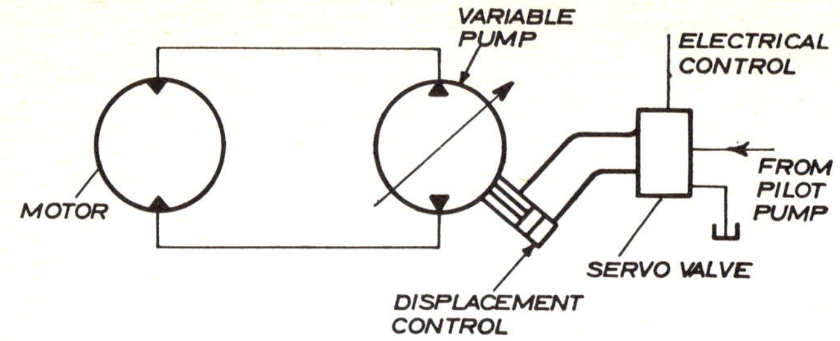

Fig. 64
Variable speed hydrostatic drive with variable displacement pump controlled by servo-valve.

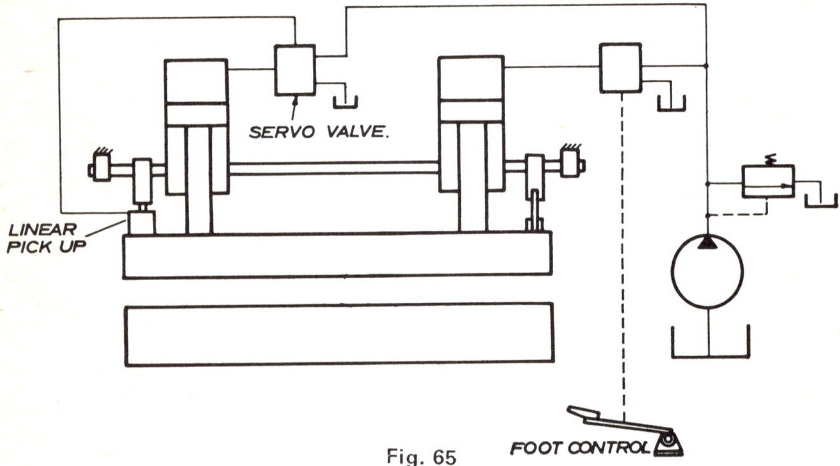

Fig. 65
Synchronising two rams as in a press brake, by electro-hydraulic servo-valve.

SPEED CONTROL WITH VARIABLE DELIVERY PUMP

The basic method of servo control of a variable delivery pump is by applying pressure to one or a pair of cylinders which determine the angle of the swashplate or eccentricity of the cam ring (Fig. 66). It is usual to take a feedback from the pump setting through a potentiometer driven from the pump adjustment. The Vickers servo valve which has been specially designed for this service, is fitted directly to the pump and the feedback is mechanical. (Fig. 67)

Although pump displacement is a more accurate flow measuring device than a flow control valve, the speed of the actuator will be affected by fluid compression which depends on pressure and leakage

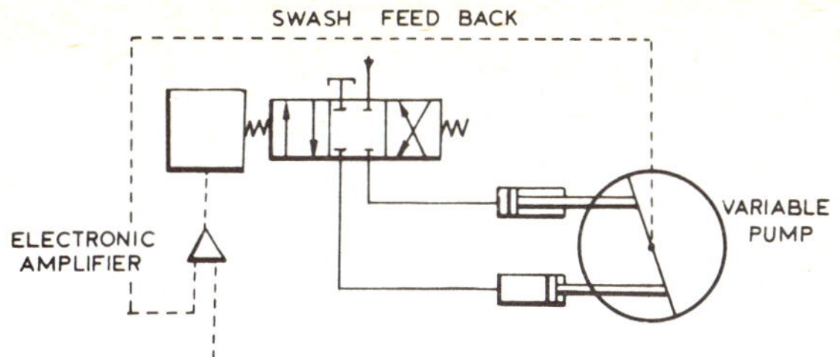

CLOSED LOOP FEED-BACK FOR PRESSURE, FLOW, POSITION OR VELOCITY CONTROLS WITH *HORSEPOWER LIMITATION OR PRESSURE COMPENSATION OVER-RIDES *

Fig. 66

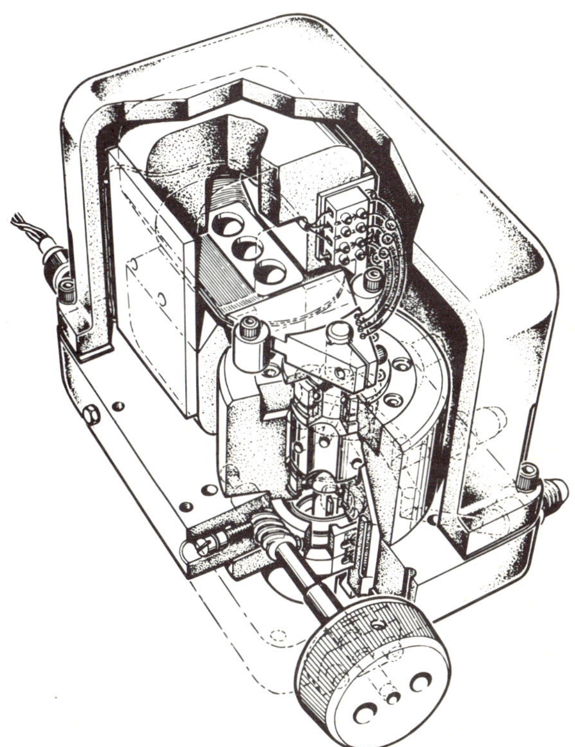

Fig.67
Vickers 'VSG' moving-coil electro-hydraulic servo valve.

or 'slip-back' which is influenced both by pressure and oil viscosity (i.e. temperature).

The function of the servo valve itself is very different from when it is acting as a flow regulator, as once the pump setting is reached, it closes. Any corrective action involves a very small movement of the valve from its neutral position. A servo-controlled pump will normally need a separate pressure supply for the servo valve. Such a system would only be justified in exceptional circumstances for drives of less than 30 horse power.

Feedback can again be given by a tachogenerator or pressure transducer and the latter can provide a convenient and accurate indication of the likely uncorrected speed change. There are many linear applications where a tachogenerator would be difficult to apply and the alternative method is sufficiently accurate.

An extremely accurate method of speed control, either in absolute terms or where two speeds have to be closely synchronized, is provided in an arrangement devised by the NEL and shown diagrammatically (Fig. 68) The basic speed is set by a frequency source as a precision speed oscillator, or picked up from another shaft which acts as the 'master . The output speed of the hydraulic motor is picked up off a disc grating by a photo-electric cell and fed to an electronic batching counter. The figure recorded on the counter is compared at regular intervals - every 15 sec - with the number of pulses in the ideal frequency in a pulse phase detector. Any difference causes an appropriate signal to be sent to the servo-valve via a servo-amplifier.

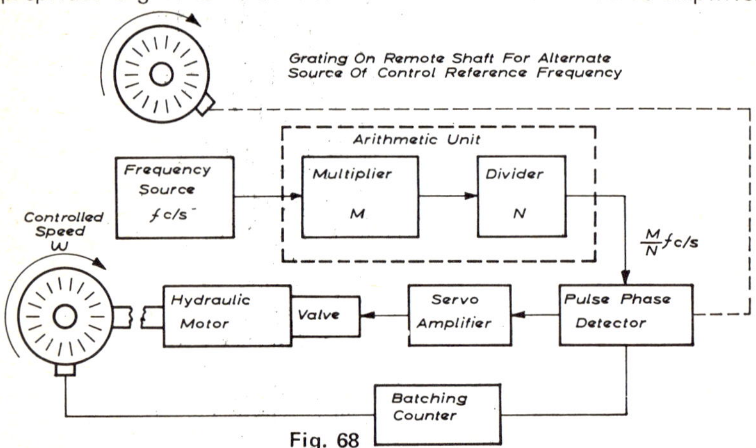

Fig. 68

Speed control by comparing actual frequency with a master frequency.

SHUNT DRIVE

A shunt hydraulic motor rotary drive is essentially a variable speed gear where the basic load is carried by gearing, usually of the epicyclic type, and a limited range of speed variation is given by a hydraulic pump/motor unit.

Chlorine gas at an ICI plant has to be drawn off a production unit at a rate relative to its generator so as to ensure a continuous negative pressure in the unit. A compressor is driven from a fixed speed 700 horse power electric motor, and this has to be regulated at a speed dependent on the signal from a process controller. The input signal and speed will vary according to the production rate but, for a given input signal, compressor speed is stabilised at that level. The compressor speed is required to be held to within $\mp 0.5\%$.

The equipment comprises an electric motor driving the compressor through a special epicyclic gearbox. A power take-off is provided at the input shaft from the electric motor driving a variable swash axial piston pump. This pump is hydrostatically coupled to a matching hydraulic motor which returns power to the epicyclic gearbox in such a way that it changes the ratio of the gearbox, thereby regulating the compressor to the required speed.

The control loop is initiated by a proportional signal from the process controller which is amplified and phase detected in a Sperry electronic modular unit; the output of the Sperry unit drives an electro-hydraulic servo valve controlling the variable swash pump.

ROLLING AND SPINNING MACHINES

The large forces involved in forming heavy metal parts by rolling or spinning may cause unequal stresses sufficient to distort the machine, but by sensing the discrepancies and using hydraulic actuators to neutralise them, precision can be achieved which would otherwise be unobtainable.

In a German spinning machine which will accept ½" thick stainless steel 16 feet dia discs the spinning operation is done by two rollers carried on arms. The force exerted by each roller can be as much as 350 tons and they must be synchronized within 0.004in.

For the large rams involved it is most economical to connect a comparatively small electro-hydraulic servo valve between the two cylinders rather than control the flow to one cylinder only as would be normal practice with smaller cylinders.

In this system the output of two resolvers, each driven by a spinning arm, is compared and the resultant electrical output is taken to the Sperry electro-hydraulic servo valve which regulates the oil flow to one or other of the valves.

In a 250-ton Bronx plate bending rolls, capable of rolling plate up to 6in thick, the roll position is also maintained to within 0.004in by a closed loop electro-hydraulic servo system. In this machine the electro-hydraulic servo valves feed the cylinders direct. The method of operation is as follows –

The height and tilt of the top roll and the position and skew angle of the bottom rolls are scarred at the rate of 250 times a second. The positions are set from the control desk and indicated on four large dials. Any discrepancy results in a signal to the servo valve to ensure the necessary corrective action.

STRIP ROLLING-MILL

Conventional rolling mills depend on the stiffness of their structure for the maintenance of accuracy, with the result that it becomes increasingly difficult to work to close limits on their gauges. Unfortunately too rigid a mill emphasises the effects of oil film, roll eccentricity and temperature.

In the Loewy system hydraulically loaded rolls are used and the thickness of the strip is controlled by measuring continuously the error in the actual roll bite and is known as the 'Constant roll gap hydraulically loaded mill'.

The mill is shown diagrammatically (Fig.69) The mill stand comprises two housings containing the chock assemblies and two hydraulic cylinders which press the two pairs of back-up chocks against each other. Incorporated in each back-up chock is the roll gap setting mechanism, the screws of which act against the bottom back-up chocks and provide the pre-stressing mechanism. The screws can be retracted so that the mill becomes a concentric hydraulically loaded mill having 'soft' characteristics.

The combination of pre-stressing and hydraulic loading enables either hydraulic or AC motors to drive the roll gap setting: as this does not have to be done under load, hydraulic cylinders protect against overload and the hydraulically actuated gauge control system is unaffected by screw-down backlash. There are also important effects on the mill design.

The automatic control system depends on the pressure in the hydraulic cylinders being regulated by a servo valve actuated by force measuring devices or load cells. One set of load cells (H) is fitted between top back-up chocks and the frame measuring the hydraulic pre-stressing force. The other set of load cells (I) is built into the bottom chocks to measure the screw load.

There are four control modes, namely:–

1. **Constant hydraulic pressure only** – useful for skin-passing, tempering and foil rolling, when roll eccentricity can produce the greatest errors.
2. **Constant hydraulic pressure with mill pre-stressed.** This method is used for accurate rolling of comparatively thick strip. The effect of roll eccentricity is minimised without sacrificing the advantages of rigidity. The effective mill spring is limited to that of the rolls and chocks.

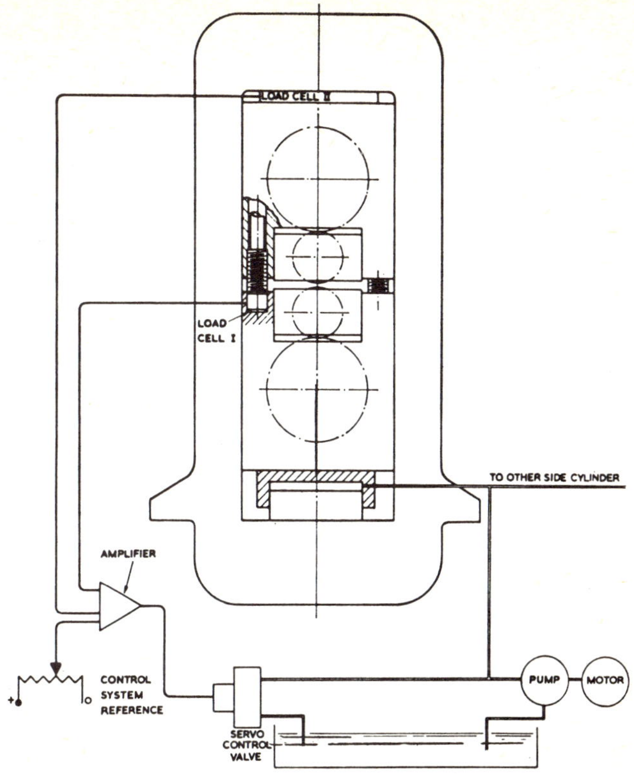

Fig. 69

Schematic diagram of Loewy hydraulically-loaded rolling mill. The rolls are forced together by the short cylinder at the bottom and separated by the pair of screws. Variations of the gap between the rolls are sensed by the load cells and compared with a reference signal to operate the servo valve.

3. **Constant screw load control.** Under this mode, gauge errors due to chock, screw and housing spring are eliminated. This gives a cheap method of gauge control most effective on narrow mills.
4. **Constant roll gap control.** This is the most sophisticated method as it constantly corrects for error as it occurs. Feedback from load cells(1) and(11) are combined and the mill becomes effectively a spring of infinite stiffness. Fig. 70. Compensation for roll deflection is obtained by a constant factor. This is achieved by an appropriate combination of feedback signals; the

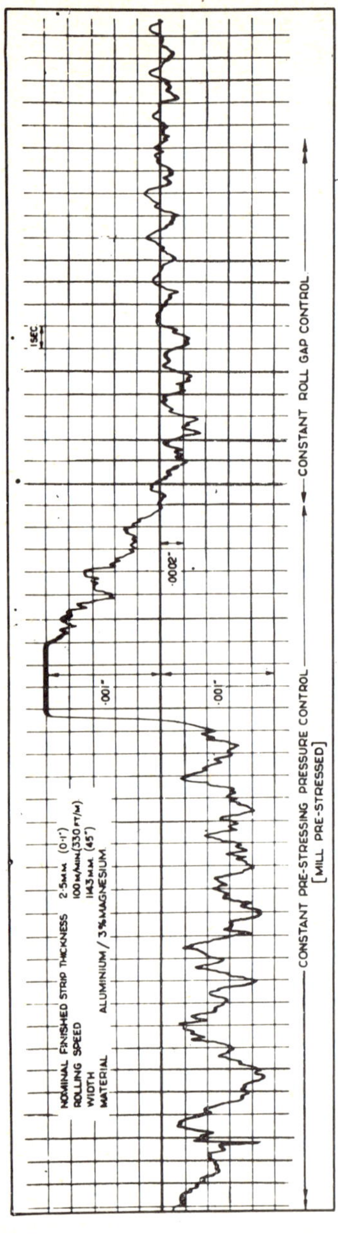

Fig. 70

Trace showing reduction of error when mill is operating on 'Constant Roll Gap Control'.

result is compared with a fixed reference signal and the appropriate gauge error detected and corrected. The time for error correction is about 30 milliseconds.

In practice, different modes are used on the same mill as the gauge is reduced. The very accurate strip produced by 'constant gap control', for instance, is furnished by mode I to eliminate the effect of roll eccentricity.

WEIGHT SAVING

Besides giving important improvement in accuracy by applying the exact load at the exact moment electro-hydraulic servo valve controlled devices open up many possibilities in making substantial savings in size and weight of heavily loaded machines.

The principles involved either –

(a) Take the forces entirely hydraulically so that the machine becomes a positioning framework only.

(b) Take part of the load by the machine frame and compensate excessive movement hydraulically. This is the technique adopted in the rolling mill described above.

An example of the first method is a hydraulically operated shear or guillotine where the upper shear is operated by a cylinder at each end. It is essential that the total force exerted should be the same whatever the position of the cut and that the shear frames should remain parallel. The two cylinders are arranged to draw the two blades together so that stresses are concentrated in the blades themselves. An unstressed reference bar is pivoted or guided so that it is always in correct alignment. It carries a linear pick-up which senses any discrepancy in the top blade alignment. The two cylinders are fed through electro-hydraulic servo valves once the low pressure pump has cut out and the pressures are added by the transducer. The build-up of pressure is controlled so that the sum does not exceed the set figure whilst at the same time the pick-up determines the relative pressures in the two cylinders. For instance, if all the cutting is done at one end then the cylinder at that end would be subjected to almost full pressure, whilst that at the other end would be correspondingly low. On the other hand, cutting a wide sheet with a sloping blade would cause a rapid realignment of pressure as the cut progressed from one end of the machine to the other.

MAINTAINING ALIGNMENT

Where heavy parts are lifted by rams through a distance of several feet the problem of lateral positioning has to be solved. Usually this is done by fitting guides sufficiently robust to withstand the forces involved. There are many cases where guides are expensive to install and are likely to be a constant source of trouble if, as often happens, the conditions are bad.

By using what, is, in effect, a copying mechanism substantial guides are avoided, these being replaced by a light straight edge and a hydraulic stabilizing cylinder. As the rams rise and fall the stabilizing cylinder responds to the pick-up signals. If necessary, substantial forces can be resisted in this way by specifying an appropriate cylinder size and pressure.

LOAD LIMITATION

With an all-hydraulic system it is usually a comparatively simple matter to ensure that the prime mover is not overloaded. In coal cutting and similar machines there are often two separate sections – an electric motor drive for the cutters and a hydrostatic propulsion unit. The difficulty here arises because the power taken by the cutter motor depends on the forward speed and the strata being cut.

By sensing the main motor current through a suitable transformer it is possible to control the output of the hydrostatic pump by using an electro-hydraulic servo valve to alter the swash angle. The object is to keep the main motor current as high as possible without overloading and to vary the forward speed accordingly. This is the system incorporated in the Muschamp thin seam coal sheaver.

POSITIONAL CONTROL

Although this example deals with a particular problem of positioning, it would be possible to use a similar technique to establish a predetermined number of positions.

In the Avery Deadweight Machine it is possible to apply gravitational forces up to 60 tons force or 60000 Kgf in increases of 1000 lbf or 500 Kgf, or as specified.

The system of operation and control of the 7150 CSB/DSB represents a complete departure from conventional testing machine practice. Control of the loading or straining of a specimen no longer depends upon the skill of the operator in operating the machine controls to provide the required programme of loading or straining. The heart of this new system is an electro-hydraulic servo mechanism which is self balancing. The machine is always under either load or displacement control and these two variables are continuously under display. This means that the load can be controlled with a resulting displacement can be controlled with a resulting load indication.

The advantage of this system is that the pattern of loading or straining can be precisely controlled and the machine responds immediately to the instruction received. Changes in specimen behaviour are automatically compensated even when non-linear. For instance, under displacement control from an extensometer, the machine will arrest a partly propagated crack and the load will adjust itself accordingly. Therefore no account need be taken of the stored energy in the frame, this being automatically dissipated by the servo valve. For the same reason the machine will hold a set of displacement or load conditions indefinitely or will apply programmes or cycles indefinitely without attention.

The straining unit of the machine embodies a double acting hydraulic cylinder and ram – (A), controlled by an electro-hydraulic servo valve – (B) fed by a pump (K). The electrical signal received by the servo valve results in a ram motion according to the magnitude and direction of the signal. The load on the specimen is measured by load cells – (C) and displayed on a linear scale – (D). The deflection of the specimen is measured either by a stroke transducer – (E) or an extensometer – (F) and displayed on the displacement scale – (G). The required load or displacement is set up by a control cursor – (H) and according to the control mode selected, either the load – (G) or the displacement – (D) is under control. Any difference between the control setting – (H) and the required load or displacement appears as an error signal across the error detector amplifier – (J), which then produces a signal which operates the servovalve – (B) – see Fig.71. Another modern testing system is shown in Fig, 72

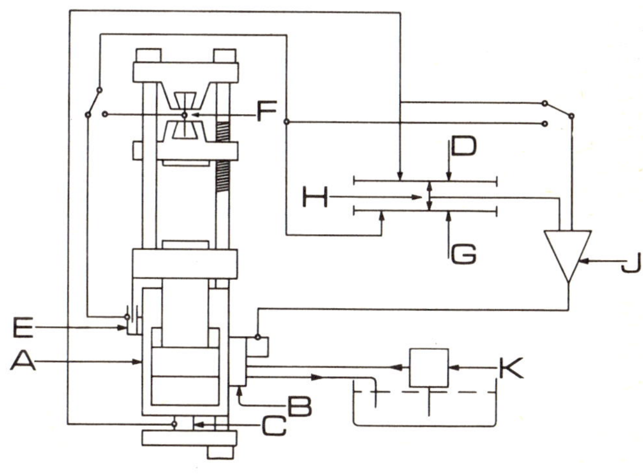

Fig. 71

NUMERICALLY CONTROLLED MACHINE TOOLS

Electro hydraulic servo valves are used for both main spindle speed control and for table positioning, the latter particularly on continuous path machines. In the Kearns numerically controlled horizontal boring

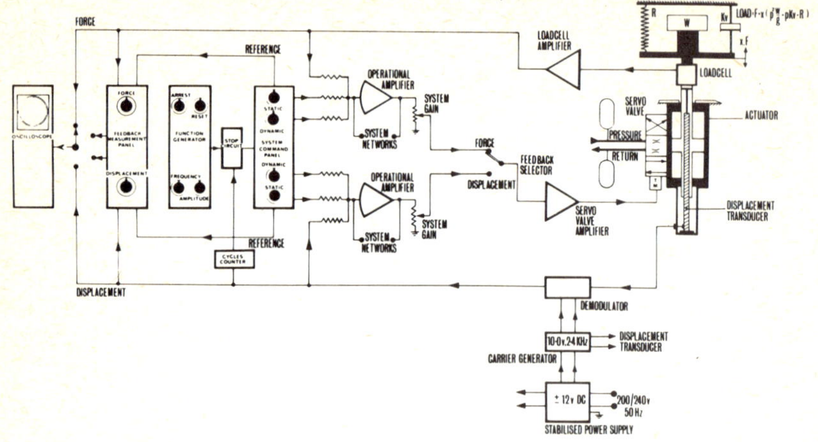

Fig. 72

Typical system diagram of Keelavite dynamic test system using either force or displacement feedback.

machine the spindle is driven hydrostatically, speeds below 300 rpm being selected by altering the pump delivery and from 300 rpm to 2000 rpm by altering the motor displacement. A voltage signal is given from the programme or set manually, which is proportional to the required speed. This is compared with a voltage from a tachogenerator driven by the spindle and any error is used to actuate the servo valve controlling the pump or motor swashplate angle.

For positioning it is the duty of an electro-hydraulic servo valve controlled hydraulic motor to rotate a recirculatory ball lead screw to conform with the signals given by the measuring system – usually optical.

In the Ferranti system the early examples had electric motors, but the introduction of hydraulic motors in 1958 was found to bring many advantages and this is now the drive used in most systems.

The actual machine control medium is magnetic tape which has been prepared on a computer. The machine has a scanning disc on which are marked a number of Archimedian spirals. These, in conjunction with an optical grating may give a pulse, via a photo-electric cell for every 0.0001in table displacement. If the pulses picked up correspond then no signal is produced; if they are out of phase it indicates that backward or forward movement is required and a corresponding signal is sent to the servo valve controlling the lead screw motor. To give the rapid acceleration and deceleration essential for the success of the system the hydraulic control is ideal.

Pump Servo Systems

In a pump-operated servo system the pump is part of the servo loop, and the working pressure is variable depending on the load. By eliminating throttling (as with valve-operated servos), efficiencies are generally higher and such systems can be used for high power outputs without overheating and power dissipation becoming a major problem. Dynamic stability may, however, be less satisfactory, and thus valve-operated servos are almost always preferred for lower power ratings.

A typical pump-operated servo system is shown in block diagram (Fig. 73)The input signal is applied directly to a servo valve controlling an actuator (or power amplifier) which in turn operates the delivery control on the variable delivery pump. The input thus directly controls the flow to the output actuator. For a positioning system the pump must be of the reversible type, so that both magnitude and

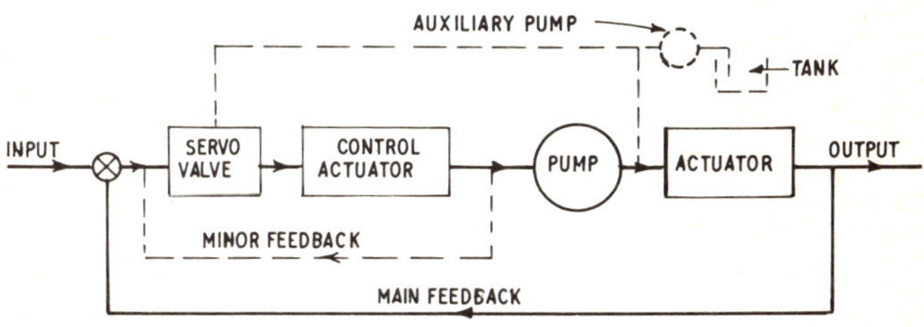

Fig.73.

direction of flow are controlled by stroke control actuator. In practice, an auxiliary pump may also be provided, both for operating the servo valve and for giving a base pressure for the pump servo.

Two feedback loops are shown. The main feedback generates an error signal in the main loop and thus provides the required proportionality. The servo valve and stroke control actuator also operate on a closed loop, the output movement of the stroke control actuator being operated by the minor feedback loop error.

Where the same set-up is employed as a velocity control system the permissible loop gain is low, particularly at low natural frequencies. This can lead to substantial errors in velocity and load. A method of overcoming this is to dispense with the minor feedback loop which will appreciably improve the steady-state performance of a velocity servo without introducing instability. This will, however, have an adverse effect on the dynamic stability and response.

A number of other types of hydraulic servo controls are shown in Fig.74.

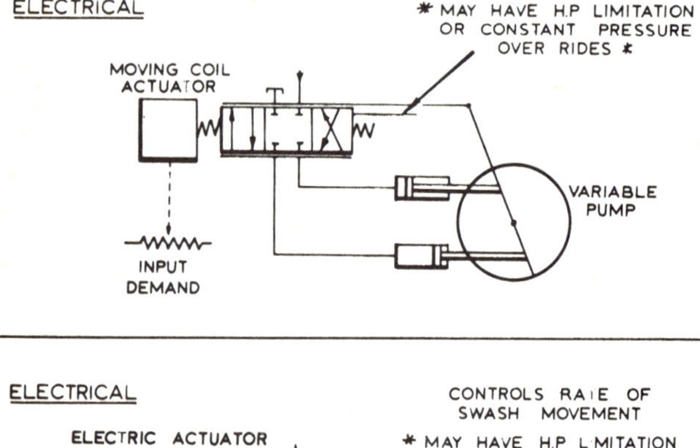

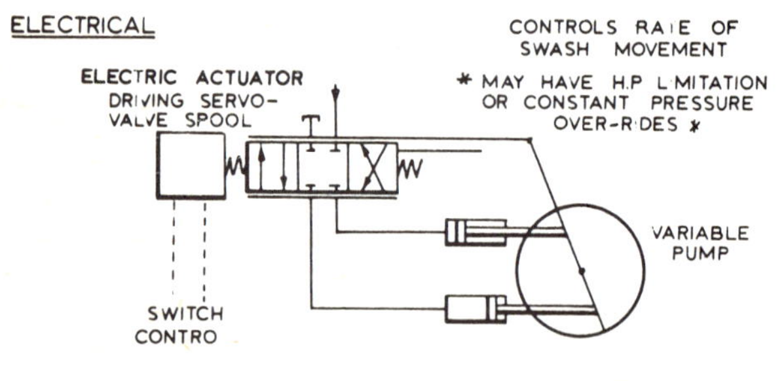

Fig.74

(Courtesy Towler Hydraulics).

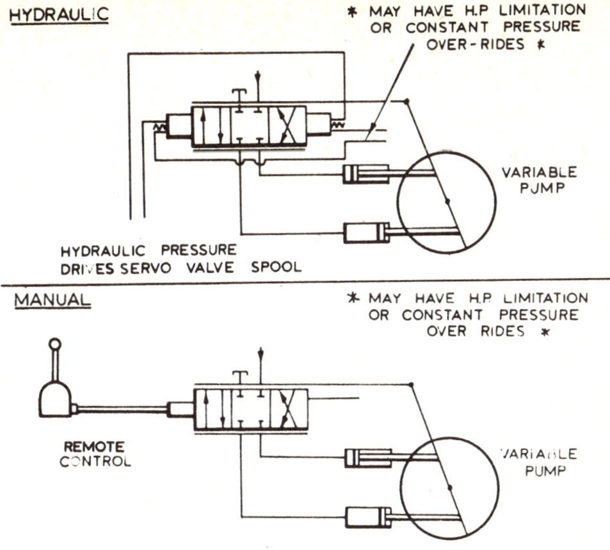

Fig. 74 cont'd.

Fig.75 is a functionaldiagram of the swashplate or tilt box positioning, the servo valve directing flow to and from the appropriate stroking cylinder. Fig.76 shows a typical pump servo valve action (Vickers) in more detail.

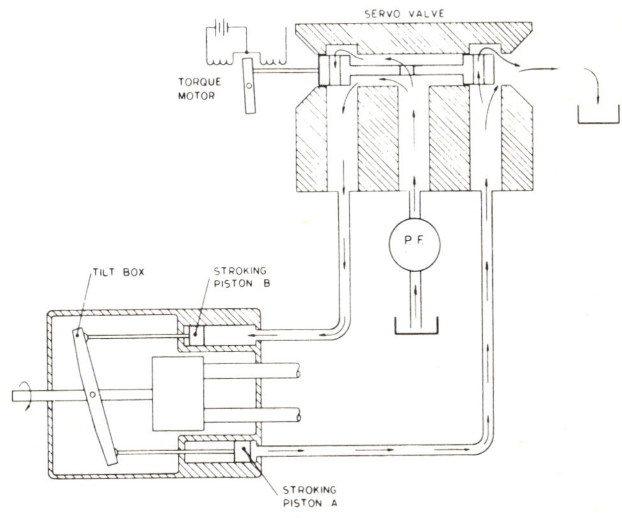

Fig. 75

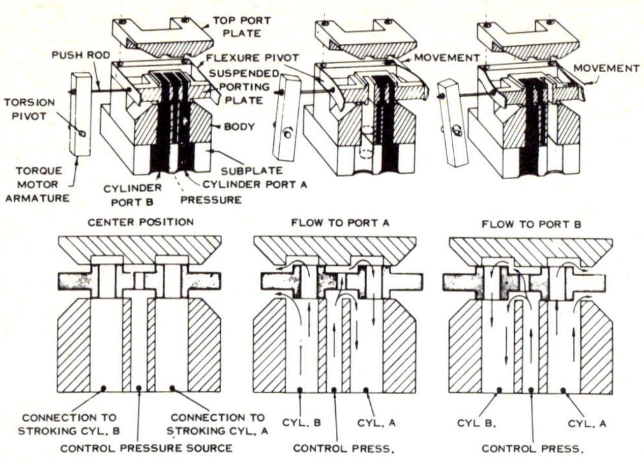

Fig. 76

The servo valve is either mounted on top of the pump or incorporated in the pump casing (e.g. see Fig.77).

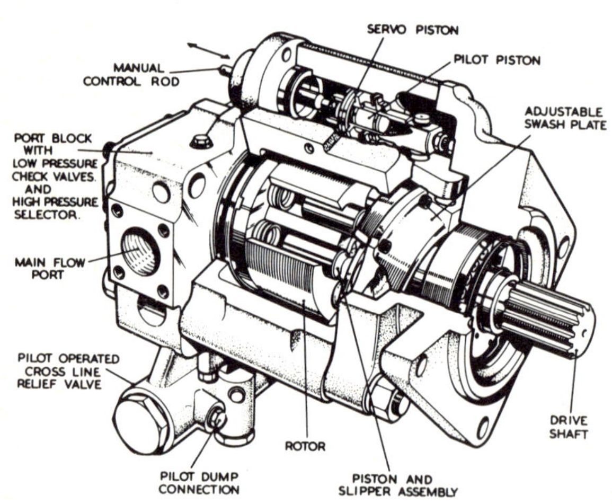

LUCAS INDUSTRIAL EQUIPMENT LTD., LIVERPOOL, ENGLAND.

Fig. 77

Digital and Pulse Controls

Servo systems so far described have been of the continuous type where the action of the controller is a continuous function of the error. Discontinuous systems may operate on simple on-off signals, or from digital or pulse inputs.

On-off or 'bang-bang' control may be appropriate where the modulating control given by a continuous servo system is not suitable, or not necessary. An elementary example is a thermostat operating as an on-off switching control and thus performing the function of a two-position controller. Equally, bang-bang control may be used where great accuracy of control is not required and the conditions are such that they might cause unpredictable instability with a continuous servo-system.

This is the method adopted for the automatic levelling of the blade of a road grader so that once the correct angle relative to the horizontal has been set the blade maintains this angle irrespective of irregularities of the ground.

A (damped) pendulum unit is mounted on the blade itself and this provides a signal which is compared, via a selsyn, with a signal from a setting scale. When a deviation occurs a signal relative to the direction of deviation is fed via an amplifier to one of two solenoid operated valves. One of these acts to raise the end of the blade, the other to lower it.

DIGITAL ACTUATION

There are also a number of actual and potential applications where the information available is presented in the form of one or more pulses, and since this is the type of information given by digital computers an increase in this form of control is bound to develop, although direct digital actuation is still a relatively new approach.

A clear distinction is drawn between a binary system and a pulse system. In the case of a binary system movement of the digital actuator takes the form of a number of discrete positions, when the controller can take the relatively simple form of a series of switches and relays, or logic elements. The system then operates under open loop conditions – although it does have an absolute datum. (Fig.78)

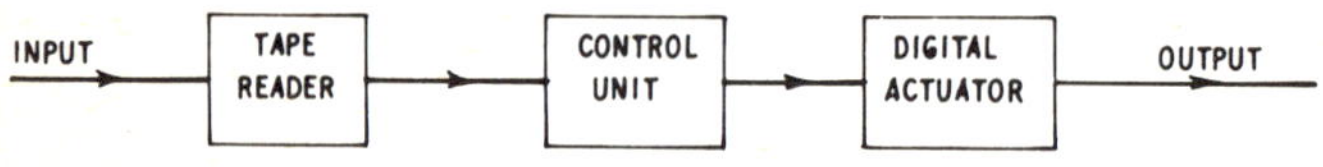

Fig. 78

In the case of a pulse system the set-up is somewhat more complex – (Fig: 79) The binary input signal is converted into a corresponding number of pulses which are then fed to an electrohydraulic pulse motor. This comprises an electro-mechanical stepper motor coupled to the input shaft of a rotary servo valve controlling a hydraulic motor, together with feedback (usually mechanical) so that

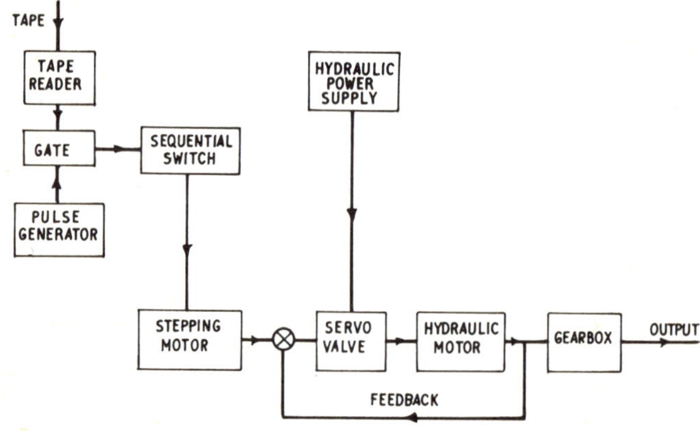

Fig. 79

the hydraulic system is operating under closed loop conditions. The application of a pulse rotates the stepper motor through a discrete angle, with a corresponding angular displacement of the servo valve input shaft. The error signal generated in the feedback loop causes the output shaft of the hydraulic motor to drive to a corresponding position. Thc numbcr of pulscs applied thus governs the position of the output shaft, whilst the speed of rotation of the shaft is governed by the pulse rate.

The resolution follows as

$$\text{error (e)} = \frac{\text{load velocity}}{60 \times \text{pulse rate (pulses per second)}}$$

where e and load velocity are in consistent units.

A typical pulse rate for a pulse system is 2000 pulses per second, whence the resolution for a load velocity of 120 inches per minute would be 0.001 inches.

Although operating on a continuous servo system a disadvantage of the pulse system is that it has no absolute datum and thus the dropping of any pulses would result in accumulative errors in positioning.

Some practical examples of digital and pulse controls are given below. A principal application at the present time is to heavy forging presses where the advantages are attractive since it is possible to produce forgings which are more accurate, and require less re-heating, than those made in the traditional way with hand control of the press.

The system used is similar to that which might be applied to any computer controlled system. In this case the number of pulses needed to reach the final ram position are predetermined and the ram, as it descends, causes pulses to be emitted which are subtracted from the predetermined number. When the result is zero, minus one, a valve is triggered which initiates ram reversal.

The problem is essentially that of reversing the press ram at a precise position relative to the anvil supporting the work, so ensuring a forging thickness which is within $\pm 1/32$ in of the required size. The speed may be 100 or more strokes a minute and the force 3,000 tons.

This system is not error operated and the ram is reversed as quickly as possible. On the press it is a repetitive positioning device but there is no reason why the final position should not be changed as often as necessary by a computer, so that it could be used as a general positioning device for heavy parts. The accuracy is dependent on the increments of the digital system.

The system to be described was developed by Towler Hydraulics and associated companies and operates as follows:–

The forging dimensions are determined by the distance between the fixed bottom and moving top tool at the moment of reversal. The position of the top tool is relative to the press frame and is continuously measured by a counting device called a digitiser which is driven by a cable attached to the ram and returned by a coil spring. When setting the press the figures representing the final tool position as indicated by the counter are recorded. When the ram is falling, the counter pulses are subtracted from those recorded until they cancel out, the return valve is triggered and the ram reverses. Correction is made for ram speed, as detected by a tachogenerator, to prevent overshooting and, by measuring the hydraulic pressure for frame stretch. The ram velocity may be 30in per sec.

Hydraulics System

This follows closely the practice for very large direct coupled presses, the principal difference being the electro-hydraulic servo valve which is of the moving coil, permanent magnet type, which can be energised from a transistor and yet passes sufficient oil in 15 milliseconds to operate the large main control valve. The speed is increased by using an initial large current pulse. The valves are used to reverse at both end of the stroke and these are arranged so that failure of either valve cannot result in over penetration and the spoiling of a valuable ingot.

The main valve is operated by the two servo valves. For bottom reversal both valves apply a 'blip' of pressure to the main valve which is held in position by a detent. If either servo valve fails the other valve is sufficient to operate the main valve. At the top of the stroke the pressure held in the main valve is normally exhausted through the two servo valves in series. If one has failed, the passage is blocked and the press is stopped.

Electrical System

The electrical system is probably unique and is based on the digital method of control, using binary arithmetic in conjunction with a transistorized logic static switching system.

A digital control system is one which controls by discrete steps, as distinct from an analogue system where there is infinite modulation.

The best accuracy which can be obtained with an analogue system under industrial conditions is about 1 in 500. There is no limit to the accuracy of the digital system but for a forging press, the accuracy corresponding to 1/32inch (0.79mm) in 128 inches (3251mm) or 1 in 4096 is sufficient.

From the foregoing it will be seen that were it not for the tremendous speeds and accelerations involved, it would be possible to use mechanical counters which operate electric contacts when the desired figures are reached. The high speeds make electronics inevitable and the methods used are based on the binary system employed for computers.

The Digitiser

The digitiser is, in effect, the counter, but is arranged to give signals to η circuits representing the power of 2 from 2° to 2^{η}.

Mechanically the digitiser consists of any number of discs geared together in the suitable ratios and rotated by a cable attached to the press crosshead. The discs have magnets covering the shaded areas, embedded in epoxy resin, and move over pick-up coils which are neutralised when a magnet is above them, so that there is no signal (represents 0). A signal represents 1.

Setting switches then allow the operator to set the forging thickness by dials to ,say, 1/32 inch (0.79mm) and this is converted into binary signals by a simple decimal-binary matrix, these being taken to the adder where they are combined with the zero height signals. The adder is a transisterised logic unit. The function of the digitiser has already been explained. The zeroing equipment works as follows:– the press is brought tool-to-tool, without a workpiece and the digitiser output is taken to η lamps, one for each power of 2 A switch below each lamp is then set 'on' if the lamp is illuminated and 'off' if it is not, the signals being taken to the logic adder so that the signals from the adder to the subtractor now represent the bottom tool height plus work thickness, or the reversing position of the tool.

Signals representing hammer speed and frame stretch are injected into the electronic system and a variable dwell of 0 – 5 seconds is provided at the top of the stroke to give time to manipulate the ingot. It is also possible to decelerate the press, which may be falling at three times the forging speed, at about 1 inch (25mm) from the bottom position. This fast approach saves time better control is given if the press is slowed down. The deceleration signal is also obtained from the digitiser.

This electronic-hydraulic system of control has shown great reliability and the electronic system is composed of a number of simple standardised units, all easily replaced by merely plugging in a new unit.

COMPUTER RESET

The system just described is limited in its accuracy by the difficulty of producing a mechanical digitiser capable of dealing with the large numbers involved. A figure of 0.001in over a distance of 36in would mean 36000 separate positions, involving a 14 figure binary number.

It is, however, possible with an error sensing computer system to use the read-out to reset a modulating control. It would be possible to reset a potentiometer-type control by a pulse motor or similar device. Similar systems are already being applied to resetting pneumatic controllers.

DIGITALLY CONTROLLED SYSTEMS

Digitally controlled systems, in addition to being extremely versatile, accurate and fast, eliminate the mechanical settings of lengths. Cut-lengths and total number pieces can be set by the operator on the operator's panel, or this information can be programmed from punched cards or tapes.

Fig. 80 illustrates a digitally controlled drive (Vickers) which has a rotary transducer mounted on the slide for measuring the differential movement between the cutting die and material. Coupled to the transducer shaft is an accurate measuring wheel which rides on the material, i.e., the transducer shaft is rotated proportionally with the linear travel of the material. The shaft of the transducer, for example, if coupled to a measuring wheel having a 12-inch circumference would rotate 1 revolution for every 12 inches of material, or 1/1200 revolution for each material increment of 0.010 inch.

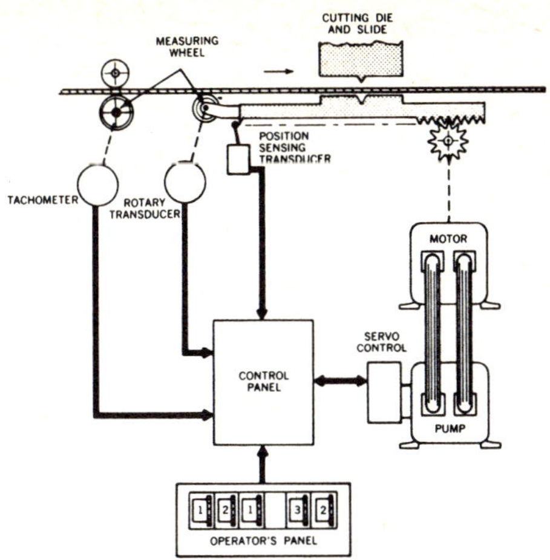

Fig. 80
Digital Position Control

As the material travels through the cutting die, the transducer supplies the control panel with signals corresponding to the number of increments the material has travelled relative to the cutting die.

These signals, in effect, are accumulated, counted and compared with the value set by the operator on the operator's panel. To obtain a cut-length, for example, of 121.32 inches, the required operator's panel setting would be 12132. This means that in order to achieve the desired length, the transducer shaft must rotate a total of 12,132 'step' being equivalent to 1/1200 revolution of the transducer shaft or 0.010 inch of linear material travel.

Just prior to the actual length of the desired cut being reached, the output of the material driven tachometer is applied to the servo pump. The tachometer output being a function of material velocity, commands the servo pump to accelerate the slide to the velocity of the material. At the same time, the rotary transducer provides a slide position feedback signal which, in effect, causes the slide and material to 'lock' at the commanded cut-length through the cutting operation.

After the cut has been completed, the output of the position sensing transducer is applied to the servo pump, causing the slide to reverse and mechanically actuate the position sensing transducer at the return position of the slide. This nulls the output of the position sensing transducer and holds the slide in position ready for the next cut.

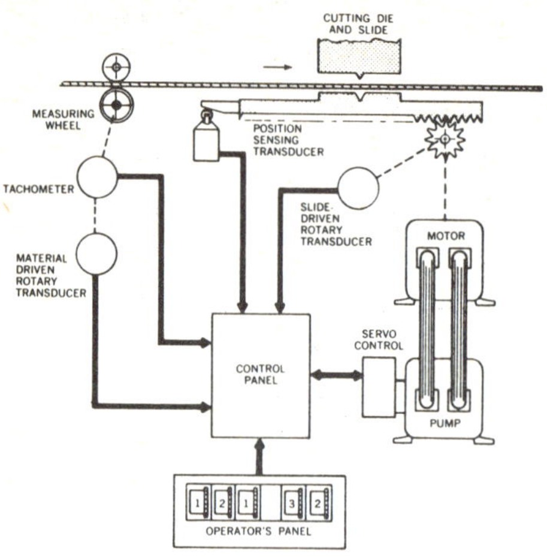

Fig. 81

Digital Position Control

Operation of the digitally controlled drive shown in Fig. 81 is similar to that of Fig. 80. The basic difference is that the rotary transducer mounted directly on the slide is replaced with two separate transducers. The resultant, however, is the same – a signal equivalent to the differential movement between the slide and material. This arrangement is effective where the direct mounting of a transducer on the slide is not practical.

As previously mentioned, the drive must accelerate and synchronize the slide with the material in the minimum possible time. Synchronization for a typical Vickers RVA digitally controlled drive occurs between 0.15 and 0.25 seconds after slide acceleration, while position error is reduced to approximately 0.020 inch after 0.20 second. Relationship of die velocity and position error with respect to time for a typical drive is shown by the oscilloscope trace facsimile, Fig. 82.

Fig.83 illustrates another Vickers RVA digital position control which is an extremely versatile, fast, and accurate control for a feed-to-stop drive. This drive uses a rotary transducer, a resolver,

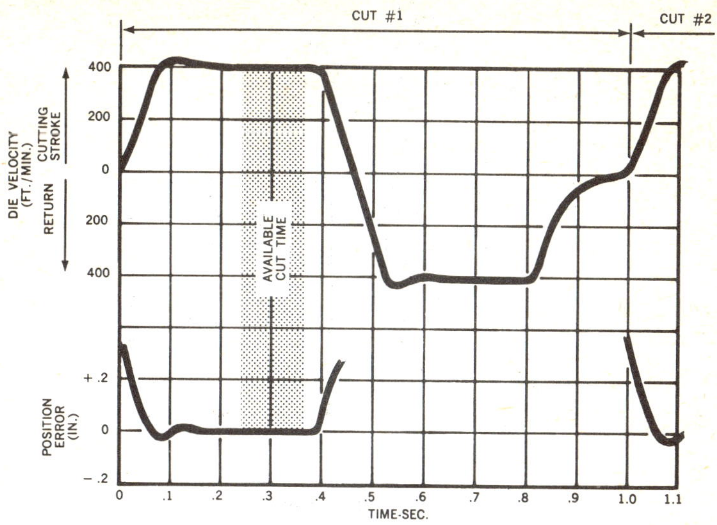

Fig. 82
Typical response curve for a Vickers RVA digitally controlled drive.

for measuring material travel. The resolver, a device similar to a synchro, produces an electrical signal that is a function of its shaft position. It can be coupled to an accurate measuring wheel riding on the material, or directly coupled to the motor using the feed rolls as the measuring wheel. In positioning systems, such transducers are used in pairs, a command and a follower unit. Errors in shaft position of the two units would be used to signal the follower drive and transducer to stay in position step with the command transducer.

In the drive shown, the command transducer is replaced by an electronic 'synthetic resolver' in the control panel. This control produces electrical signals exactly the same as that coming from a rotating command resolver, and the feed roll drive is made to follow this command. Furthermore, the amount of 'equivalent rotation' of this synthetic resolver can be set digitally by the operator at the operator's panel.

For a typical example, assume that (a) measuring wheel circumference is 10 inches. (b) incremental settings as small as 0.010 inch

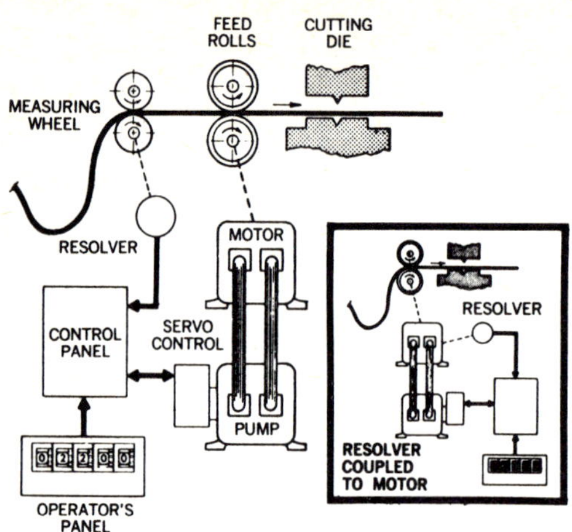

Fig. 83

Digital Position Control

(1/1000 of a revolution) are desired. Then, if a measurement of 22 inches is desired, the measuring wheel must rotate 2.2 revolutions. The required command setting of the operator's panel would be 022.00 (2.2 revolutions). This actually means that the drive must rotate the feed rolls an equivalent of 2200 steps, each 0.010 inches long. When the 'start' signal is initiated, the 'synthetic resolver' begins to provide a signal equivalent to a rotating resolver at the desired rate of feed roll rotation. This signal is compared with the resolver signal on the feed rolls. The difference between the signals causes the pump to drive the motor in step with the synthetic resolver.

When the resolver reaches equivalent rotation several inches from that representing the desired length, the control begins to reduce pump output and thereby, motor speed until the motor is brought to a complete stop at a position equivalent to 2200 steps of 0.010 inch, or a total of 22 inches.

This system can be instantly reset and is extremely fast, smooth, and accurate. Speeds up to 1200 fpm and accuracies ± 0.30 inch are obtainable. Stopping time from 1200 fpm can be less than 1/10 sec. – see Fig. 84.

All the electronic controls are the solid-state type designed specifically for reliable operation in industrial environments. The

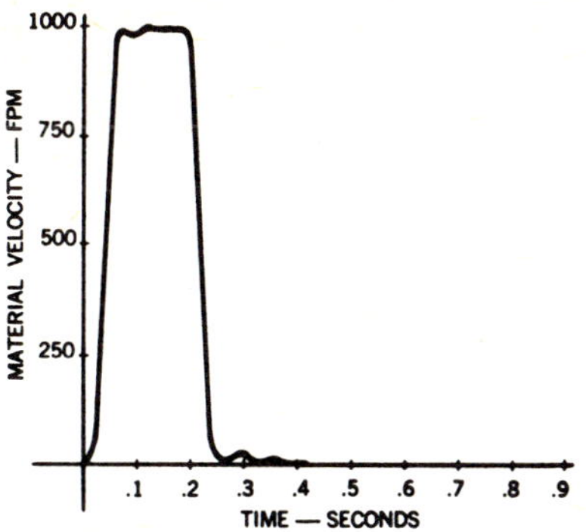

TYPICAL MATERIAL SPEED CURVE FOR 3-FT FEED LENGTH

Fig. 84

digital controls are the integrated circuit type which have proved to be the ultimate for reliable and maintenance-free operation.

PULSE MOTORS

Pulse motors provide a means of controlling position by using electrical pulses to operate a special motor, each pulse causing the motor to rotate through a definite angle. Because of the low power of such motors they are usually connected directly to a hydraulic motor to act as an amplifier. A typical motor might rotate 2° 15′ for each pulse, the direction depending on the terminal to which the pulse is applied. The maximum pulsation rate might be as much as 8000 pulses per second.

The hydraulic motor is controlled through a pintle valve which opens when the electric motor rotates and closes when this hydraulic motor is in phase with it.

As long as the hydraulic motor is not overloaded it will keep in step with the pulses. For operating a machine tool these may be generated by a magnetic tape and for many purposes it is possible to obtain sufficient accuracy by rotating recirculatory ball lead screws through the required angle.

Valve Characteristics

Basis of any hydraulic servo system is that control of the actuator movement or output is initiated by varying either the flow or pressure in the hydraulic system. For *position* control (of the actuator), either the flow rate or pressure can be controlled, the former normally being preferred as displacement/pressure proportionality deteriorates with increasing load. For *velocity* control systems, manipulation of flow rate is employed. Simple position control can be achieved merely by varying the direction of flow.

In a valve-controlled system the source of hydraulic power (the pump) is external, valve movement manipulating the flow, or pressure, as appropriate. In a pump-controlled system a variable-delivery pump feeds the actuator direct, the input being applied to the variable delivery control of the pump (usually through a valve-amplifier unit).

A three-way valve (Fig.85) is used to control a single-acting hydraulic cylinder (with spring return); or a double-acting cylinder with differential piston areas (differential jack), giving it symmetrical characteristics. A four-way valve (Fig.86) is used to control a double-rod or symmetrical jack. It should be noted that in both cases the jack is hydraulically locked with the valve in the neutral position. Control valves can be of various types, i.e. rotary cocks, sliding or spool valves, or poppet valves. Spool valves are almost invariably preferred for servo applications, largely because of their low operating loads. Rotary cocks require large operating forces and have high leakage rates (especially at high pressures). Poppet valves are generally excluded because they will normally seal only against one direction of flow.

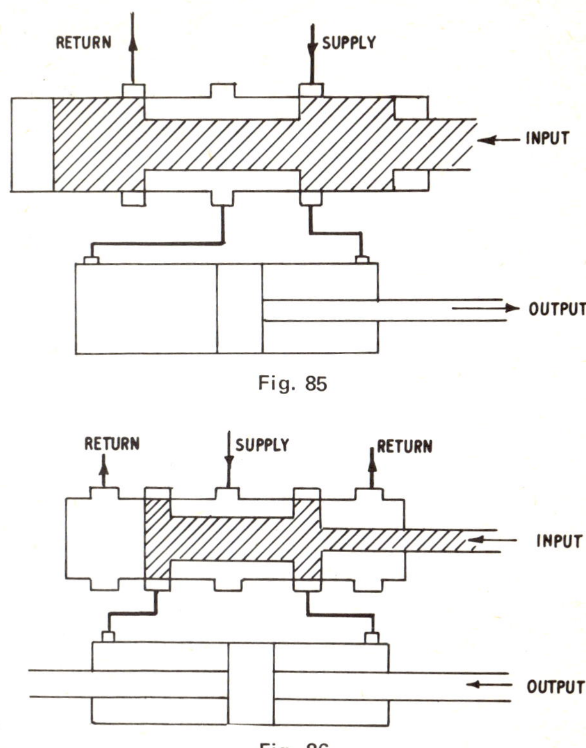

Fig. 85

Fig. 86

A feature which may be incorporated in the valve design is the provision of a central by-pass connection between the pump inlet and outlet, enabling the pump to be off-loaded when the valve is in the neutral position. This is known as an 'open centre' system. The open centre connection is closed as soon as the valve is moved in one direction or the other. The same condition is provided if the valves have underlap.

Valve detail design is concerned primarily with establishing optimum values of underlap (or overlap), where applicable; and obtaining complete static balance of hydraulic forces parallel and normal to the plunger. Friction can normally be reduced by minimising the contact area between spool and bore of the valve, but further methods may be applied, such as the application of dither or a rotary motion to the valve plunger, grooving or shaping of the spools, etc.

Under dynamic conditions, i.e. when the valve is pressurised with flow, a disturbing force is present due to the change of momentum of the fluid being throttled by the valve. This takes the form of a reduction of pressure on the valve spool at the controlling edge where the flow rate is high, producing unbalancing force on the spool, known as the Bernoulli force. This is directly proportional to the valve orifice area, or flow; and to the square root of the pressure drop across the valve, viz

Bernoulli Force $= KQ\sqrt{\Delta P}$

where Q is the flow rate
P = pressure drop across the particular controlling edge
K is the valve constant
= 0.0045 for a three-way valve
= 0.0065 for a four-way valve
With typical oil fluid of SG = 0.83
(when the Bernoulli force is in pounds; flow rate in cu.in/sec; and pressure drop in psi).

Bernoulli forces can be reduced, or even eliminated, by special shaping of the spool, but since the forces will be variable with different flow rates (i.e. valve openings), full compensation is not possible in this way. Where a valve is to be operated under relatively low inputs an alternative method of offsetting the Bernoulli forces is to provide amplification of the input signal – e.g. as in the case of electro-hydraulic servo valves. It should also be mentioned that in the presence of backlash, Bernoulli forces are capable of setting up a high frequency chatter of the valve, with consequent loss of stability.

Flow characteristics

The flow characteristics of a three-way valve are detailed in Fig. 87. The flow rate is given by:–

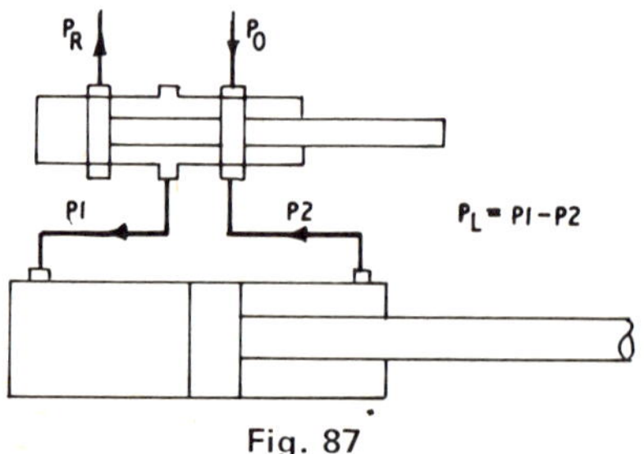

Fig. 87

$$Q = k\,a\sqrt{\frac{Po}{2} - \Delta P_L}$$

where a = area of orifice

k = valve constant = 90* for rectangular orifices

= 100* for round orifices

= 127* for annular orifices

(* for Q in cu.in/sec, pressure in psi and a in sq.in.)

The pressure drop across the valve is given by

$$\Delta P_V = \frac{Po}{2} - \Delta P_L$$

The flow characteristics of a four-way valve are detailed in Fig.88. The flow rate is given by:–

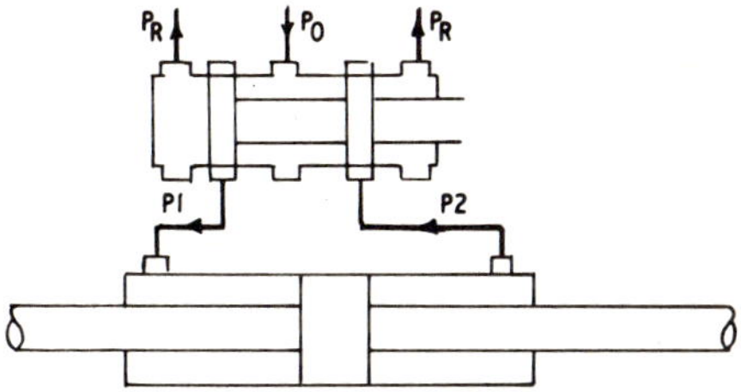

Fig. 88

$$Q = k\,a\sqrt{Po - \Delta P_L}$$

where k = 90* for rectangular orifices

= 70* for round orifices

(*for Q in cu.in/sec, pressure in psi and a in sq.in.)

The pressure drop across the valve is given by

$$\Delta P_V = Po - \Delta P_L$$

Power output

The useful power output from a valve is given by

$$W = q\,\Delta P_L$$

where ΔP_L is the pressure drop across the load

This can be rendered in the form

$$W = k_v \, a \, \Delta P_L \sqrt{P_o - \Delta P_L}$$

where K_v is the valve constant
a is the orifice area

This gives a maximum value when $\Delta P_L / P = 2/3$, viz:–

$$W_{max} = K_v \, a \frac{2}{3\sqrt{3}} P_o^{3/2}$$

or
$$\frac{W}{W_{max}} = \frac{3\sqrt{3}}{2} P_r \sqrt{1 - P_r}$$

$$\text{where } P_r = \frac{\Delta P_L}{P_o}$$

Typical pressure–flow characteristics for a four-way valve are shown in (Fig.89)

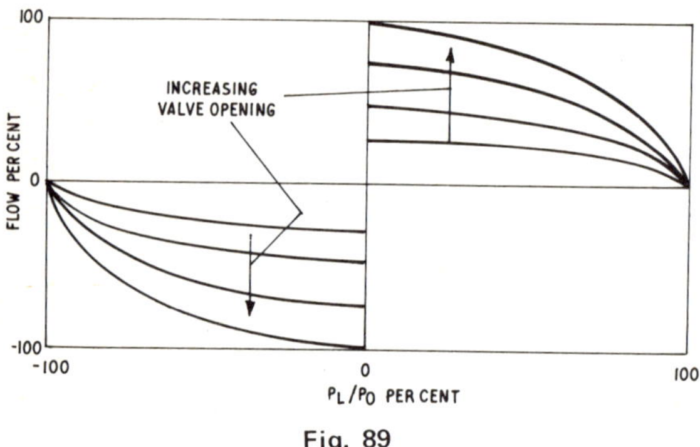

Fig. 89

Valve Lap

Flow displacement characteristics of throttle valves are largely governed by the lap conditions, as defined by Fig.90 With zero lap the valve lands are equal in width to the orifice width. Wider lands produce overlap; and narrower lands underlap. The effect of overlap is to introduce a dead zone over the lap travel, or zero slope of the flow-displacement curve. Underlap has the effect of substantially increasing the slope of the flow-displacement curve, or 'gain' over the lap travel. These characteristics are shown graphically (Fig.91) for rectangular orifices.

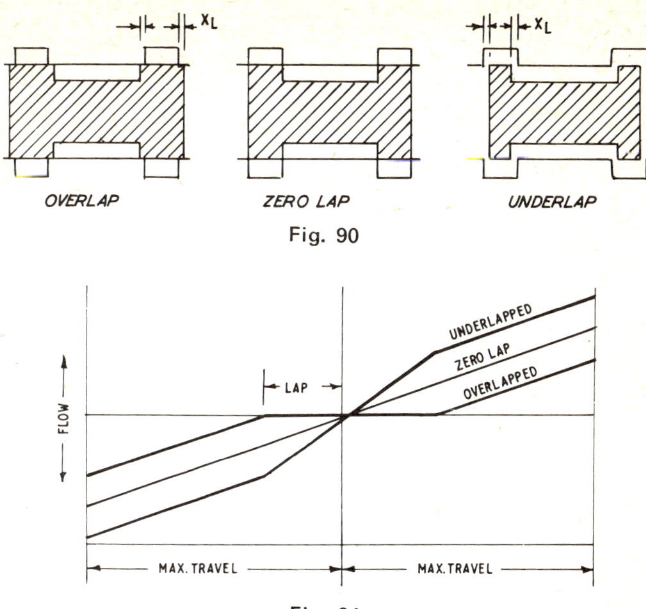

Fig. 90

Fig. 91

In the case of a valve with underlap, valve travel and valve opening differ – the former being the actual displacement of the spool from its neutral position and the latter the actual opening of the associated orifice in the direction of travel. Within the underlap region (i.e. where Y is less than X_L) the flow pattern can be analysed in terms of an equivalent Wheatstone bridge network (Fig. 92) The flow through the load is then given by:–

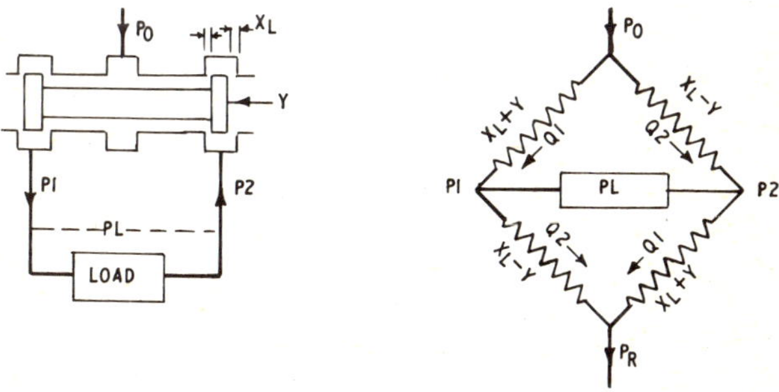

Fig. 92

$$Q = k\,w \left[(X_L + Y)\sqrt{P_o - \Delta P_L} - (X_L - Y)\sqrt{P_o - \Delta P_L} \right]$$

where w = width of orifice
X_L = underlap, as defined in Fig. 92
Y = valve travel

The valve gain, which is the slope of the flow-displacement curve, follows by differentiation:–

$$\text{valve gain } (K_v) = k\,w \left[\sqrt{P_o - \Delta P_L} + \sqrt{P_o + \Delta P_L} \right]$$

Outside the underlap region only two arms of the bridge are involved (Fig. 93) when the flow is given by

$$Q = k\,w_y\,\sqrt{P_o - \Delta P_L}$$

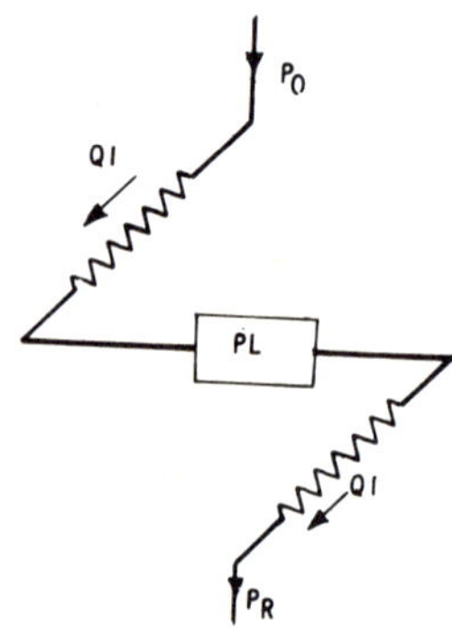

Fig. 93

The gain follows as

$$K_v = k\,w\sqrt{P_o - \Delta P_L}$$

i.e. outside the underlap region (when Y is greater than X_L) the valve characteristics of an underlapped valve are thus identical to those of a valve with no lap with an identical opening.

The other significant parameter is the pressure factor (P_v) which is defined as the flow per unit load pressure at constant valve displacement, which can be obtained by differentiating the flow with respect to the pressure drop across the load ($P_1 - P_2$). For an ideal valve (no overlap or underlap, this is given by

$$P_V = -\frac{kwy}{2}\left[\frac{1}{\sqrt{P_o - \Delta P_L}}\right]$$

For a valve with underlap this becomes (within the underlap region)

$$P_V = -\frac{kw}{2}\left[\frac{X_L + Y}{\sqrt{P_o - \Delta P_L}} + \frac{X_L - Y}{\sqrt{P_o + \Delta P_L}}\right]$$

The static output stiffness of the valve is also significant, defined as the load pressure per unit valve displacement at zero flow and generally called the valve output stiffness (S_V). For the ideal valve with rectangular orifices:

$$S_V = -\frac{2(P_o - \Delta P_L)}{y}$$

For a valve with underlap and rectangular orifices:

$$S_V = -2\left[\frac{\sqrt{P_o - \Delta P_L} + \sqrt{P_o + \Delta P_L}}{\dfrac{X_L + Y}{\sqrt{P_o - \Delta P_L}} + \dfrac{X_L - Y}{\sqrt{P_o + \Delta P_L}}}\right]$$

It is a general characteristic of an ideal valve at its neutral position that the pressure factor is zero, yielding infinite valve output stiffness at this position. With a valve with underlap, however, the static output stiffness has a finite value with the valve in its neutral position.

Valve characteristics may also be expressed in non-dimensional form, introducing valve travel in terms of a ratio $Y_r = Y/X_m$ (where Y is the travel and X_m the maximum valve opening); and underlap expressed as a ratio of the maximum valve opening, i.e. $U = x_L/x_m$ The following equations then apply:–

	Valve with no lap	Valve with underlap
Valve gain	$\sqrt{1-P}$	$\sqrt{1-P} + \sqrt{1+P}$
Pressure Factor	$\dfrac{X}{2\sqrt{1-P}}$	$-\tfrac{1}{2}\left[\dfrac{U+Y_r}{\sqrt{1-P}} + \dfrac{U-Y_r}{\sqrt{1+P}}\right]$
Valve output stiffness	$-\dfrac{2}{X}.$	$-2\left[\dfrac{\sqrt{1-P}+\sqrt{1+P}}{\dfrac{U+Y_r}{\sqrt{1-P}} + \dfrac{U-Y_r}{\sqrt{1+P}}}\right]$

where $P = \dfrac{\Delta P_L}{P_O}$ or the non-dimensional load pressure

Corresponding parameters for valves with round orifices are:–

	Valve with no lap	Valve with underlap
Flow	$\dfrac{16}{3\pi}\left[\sqrt{1-P} + x\right]$	$\dfrac{16}{3\pi}\left[(U+Y_r)^{1.5}\sqrt{1-P} - (U-Y_r)^{1.5}\sqrt{1+P}\right]$
Valve gain:	$\dfrac{8}{\pi}\sqrt{X(1-P)}$	$\dfrac{8}{\pi}\left[\sqrt{(U+Y_r)(1-P)} + \sqrt{(U-Y_r)(1+P)}\right]$
Pressure Factor;	$-\dfrac{8}{3\pi}.\dfrac{X^{1.5}}{\sqrt{1-P}}$	$-\dfrac{8}{3\pi}\left[\dfrac{(U+Y_r)^{1.5}}{\sqrt{1-P}} + \dfrac{(U-Y_r)^{1.5}}{\sqrt{1+P}}\right]$
Output Stiffness:	$\dfrac{3(1-P)}{X}$	$-\dfrac{3\left[\sqrt{(U+Y_r)(1-P)} + \sqrt{(U-Y_r)(1+P)}\right]}{\dfrac{(U+Y_r)^{1.5}}{\sqrt{1-P}} \qquad \dfrac{(U+Y_r)^{1.5}}{\sqrt{1+P}}}$

Output Compliance

Output compliance is defined as the reciprocal of valve output stiffness. A comparison between ideal and underlapped valve, with both rectangular and round orifices, is shown (Fig. 94).

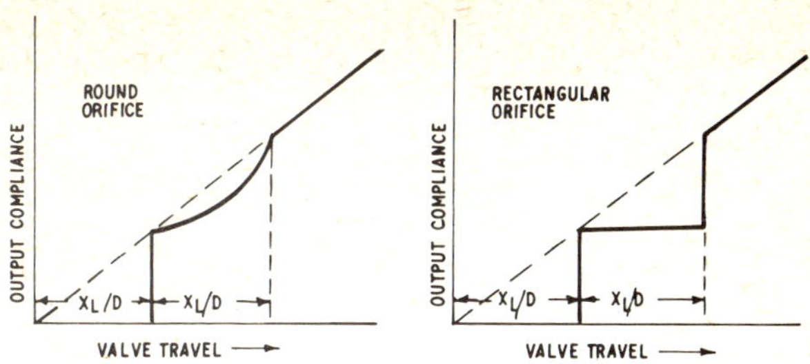

Fig. 94

Linearity

Linearity is normally taken to refer to the *deviation* from linearity of the valve gain (or proportionality to orifice width). More specifically, it is best defined in terms of the slopes of radial boundary lines to the flow-travel curve, drawn from the point of maximum opening flow (Fig. 95).

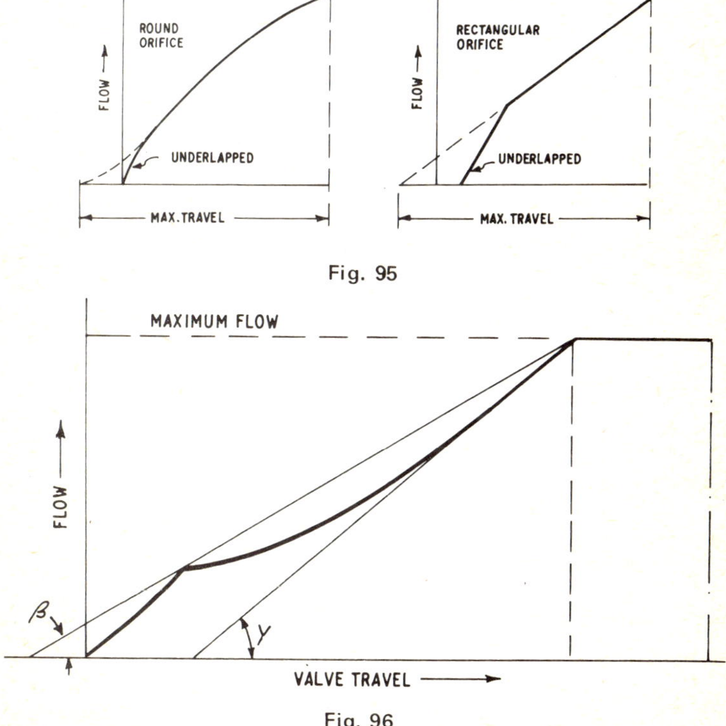

Fig. 95

Fig. 96

$$\text{Then, linearity} = \frac{\tan \alpha - \tan \lambda}{\tan \alpha + \tan \lambda}$$

Linearity is governed by the total leakage through the valve, which is appreciably modified by the introduction of underlap. It is a general characteristic that the introduction of underlap and a consequent increase in quiescent leakage will result in reduced linearity with valves with rectangular orifices, but improves the linearity of valves with round orifices (Fig. 96). In the latter case there is an optimum value of valve opening of approximately two-thirds the maximum opening, departure from linearity increasing rapidly once this opening is exceeded.

System Stability

Most of the theory of servo-mechanism is concerned with their stability, or rather the lack of it. When steam engine governors were first introduced it was soon found that there was a definite limit to the sensitivity of a simple governor. Any attempt to go beyond this results in 'hunting' as the slightest change of speed results in the valve being opened or closed far too much.

It will be appreciated that the extremely sensitive servo valves now in use can readily be affected by the characteristics of the system to which they are fitted. Unfortunately, whilst the theoretical considerations can be laid down, and indeed have been and at great length by many authors, the ultimate test on any new system is usually a practical one. In any case a glance through any book on the theory of control or servo-mechanisms will show that it is not something to be studied lightly and is essentially a subject for specialists,

PRACTICAL VARIATIONS

It is very difficult to fit servo-mechanisms into precise classes and inherent characteristics which may be important in one type may be so masked in another that for all practical purposes they are non-existent. Take, for example, steering gears: a vehicle power steering is undoubtedly a servo-mechanism as the driver's (almost) effortless input is faithfully followed by the change in direction of the steered wheels. One would not expect instability on a servo-mechanism of this kind and one reason is that the driver himself acts as a servo-mechanism and injects a stability factor into what is, in effect, the first stage of the system.

On the other hand, a completely automatic steering system, such as that operated by a gyroscope for steering a preset course, or a photocell for following a line, does not naturally possess the power of anticipation and judgment of which some of the higher animals are capable, and must therefore have its equivalent built into the system. Fortunately, this can be done on most systems where control is required. Compared with the human operator it may have some failings, mainly in dealing with unexpected emergencies. On the other hand, it does not suffer from fatigue, does just what it is told, and may be several hundred times as fast in reacting to variations as the most alert operator.

LOAD SERVO SYSTEMS

Although all servo systems have to deal with loads, only in a proportion is the load a dominant factor. Most of the pre 1939 servo systems (if steering gears are excepted) were concerned with controlling a process at a given temperature or an engine or turbine at a given speed, whilst the load varied from zero to maximum. Sometimes, where the storage capacity is comparatively large on on/off or bang-bang control such as the ordinary thermostat will give sufficiently accurate control, but if the storage capacity is relatively small then a condition of instability can easily arise.

Another domestic controller which sometimes shows that it is, despite its simplicity, subject to the laws which govern controllers, is the ball valve. It is, of course, a modulating control and the margin between 'off' and slightly 'on' is small.

On large tanks it can also show instability – identifiable by its irritating noise, caused either by ripples on the surface of the water or shocks in the supply pipe due to the sudden closing of a tap. The cure is to increase the mass or inertia and as this usually causes the float to sink deeper, it also increases the damping effect of the water.

PNEUMATIC PROCESS CONTROLLERS

Pneumatic controllers have been intensively developed during the last forty years or so and are still the backbone of the control systems for continuous processes. They are probably the easiest subject with which to demonstrate the principles of control and for this reason are treated in a separate chapter.

The principal types of control with which the controllers are intended to deal and which must form part of any automation system intended to keep conditions constant despite load changes and load inertia, are –

Proportional. The controller valve tends to fall with increase of load and rise when the load is removed, although in mechanical systems any attempt to reduce the saving beyond a certain figure may cause hunting. On the other hand it is sometimes possible to minimise the tendency to hunt so that it is of no practical importance. If it is known what the lowest load will be, this might be taken care of by a fixed input and the controlled input would then only have to supply the balance.

Derivative. This section of a controller reacts to the rate of change of output and helps to prevent overshooting. On a large system it may take an appreciable time for an alteration in input to be reflected in the system itself. It tends to smooth out other swings inevitable with crude proportional control by introducing a time lag.

Integral or Reset The change of value with load inevitable with the proportional and derivative control may not always be acceptable. Integral control will tend to correct the error at a rate which will depend on the particular system. It will not, however, make up for losses or gains.

A frequency controlled generator which must make exactly 4320000 revolutions every 24 hours will require an additional control which could be manual.

FEEDBACK

In the process controllers mentioned above, feedback is provided by the measuring instrument, thermometer, pressure element, flow meter etc. In a speed governor it will be taken from the position of the spinning weights or in diesel engine governors from the pressure developed by a pump discharging through an orifice. In present practice the feedback will probably be electrical and the current generated by a tachogenerator at the output end will be compared in the electrical equivalent (current) representing the input speed; or if the relative speeds are very important, the input and output will each derive signals which will produce a connecting signal capable of defining the error.

Whatever system is used it does not alter the fact that the momentary output speed is affected by load inertia, springiness and friction and that there is a time delay before any feedback signal can take effect. We are, in fact, back to the basic problems of control.

It is these four factors which must be taken account of in working out a suitable method of control.

ERROR ACTUATION

All closed loop servomechanisms are error actuated, in the sense that no action is initiated unless there is an error between the input valve and the valve signalled by the feedback. There is usually a threshold valve or dead band when the magnitude of the error is not sufficient to overcome the resistance of the valve to change or the valve requires a certain minimum movement to change its output.

'Error' can be reduced by various expedients, not all of them applicable to servo valve systems. One way is by detecting the likely cause of an error before it has had time to overcome the inertia of the system sufficiently to produce a detectable effect. This method is used for electrical generator systems where it is a comparatively simple matter to detect changes in electrical demand by an ammeter and feed an appropriate anticipatory signal to the governor system.

Another method involved taking a base load by a constant energy device and using the variable device for timing only. For example, a hydraulic motor might be required to have an accurate output over a limited speed-range. If a fixed delivery pump is used to give the displacement need for the lowest speed and having in parallel with it a comparatively small variable displacement pump, only the latter would be servo controlled. Incidentally, the complete system might be cheaper than one needing a full size variable delivery pump.

POSITIONAL CONTROL

This kind of control is used for such applications as gun laying—for which purpose it has probably had more attention paid to it than for any other—steering and machine tool control. Although loads are involved they are incidental and the principal object is to get the gun, or machine tool, in the desired position at the right moment. Operation may be continuous or spasmodic depending on the application.

A typical servomechanism for positioning (Fig.97) consists of the usual valve and double acting cylinder, but the total load has been shown as it would appear in practice—an inertia effect, friction, springiness and a resisting force. The force might be a major factor or, as in many machine applications, negligible.

If it is desired to reach a predetermined position and rest there, the error will depend on friction and springiness and may oscillate several times before settling down to rest. Should the motion be reciprocating, much will depend on the speed and relation between load, inertia etc., and damping, friction etc. which can affect appreciably the feedback and cause it to get out of phase with the input.

It will be noted that another variable, backlash, may also be introduced. This is one of the most awkward factors to cater for and often depends on the care with which the various parts are made and assembled.

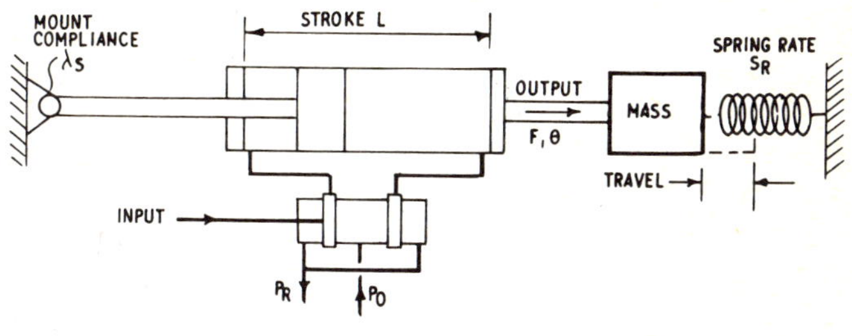

Fig. 97

In practice, arrangements vary so much that it would not be possible ot pick out a particular application. Each case must be considered on its merits, bearing in mind the factors mentioned—accuracy of manufacture, springiness, inertia, resistive force, friction and damping—although certain generalisations apply from which specific methods of analysis may be derived.

BACKLASH

Backlash in itself need not be a major problem since in a critical application it should be obligatory that backlash be eliminated. It may, however, subsequently develop through wear on mal-alignment, particularly at actuator attachment points. Where backlash is inevitably present, methods of treatment range from the use of somewhat complex phase advance mechanisms to the simple fitting of a hydraulic damper between the input and output circuits.

Component elasticity

Deformations of the individual components in the system under pressure and load will have a similar, but less marked, effect to fluid elasticity. The two may be considered together as overall elasticity, as affecting the stability if the system, although the possible treatment differs. Thus in a critical design, where physical deflections of the component members may be inevitable, it is usually possible to arrange the linkages that any displacement of the servo-unit due to deflections do not produce any movement of the control valve. The mechanical system is then 'neutrally stable' with regard to deflections. Alternatively, the system may be rendered 'positively stable' by arranging the linkages so that structural deflection feedbacks are negative, or have a valve-closing tendency. This was commonly done on earlier aircraft power control systems (where structural members have limited rigidity),before the advent of more sophisticated systems.

Irreversibility

In particular servo-system applications the closest possible approximation to an infinitely rigid mechanism may be desirable, in order to eliminate flutter or chatter. In this case the kinematic stability of the system must be studied in detail, and particular attention given also to both the backlash between individual members comprising the complete system and to the physical rigidity of individual members.

The problem can often be relieved by using an irreversible actuator, such as a low efficiency screw jack driven by a hydraulic motor. This will then give high rigidity within the limits of backlash which may be present, and physical rigidity of the system, provided there is a high degree of rigidity in the servo-motor itself.

In the case of a linear actuator (i.e. a hydraulic cylinder) the problem of rigidity and irreversibility is aggrevated at high loads by the inherent compressibility of the fluid. One straightforward solution which may be adopted in a high pressure system is to employ a differential area cylinder which is subject to a continuous constant pressure on the smaller piston area, so that in the neutral position fluid pressure is applied to each side of the piston. A high degree of rigidity can be achieved in this manner using a 'loading' pressure of the order of 500 psi.

Compliance

The springiness or elasticity of the system is generally referred to as compliance, when the total compliance (λ) is given by

$$\lambda = \lambda_f + \lambda_s$$

where λ_f = fluid compliance

λ_s = component elasticity, including mounts, which may usually be negligable or rendered negligible, as shown in the previous paragraph.

In the case of a linear actuator the fluid compliance can be derived in terms of the piston area (A), stroke (L) and the bulk modulus of the fluid (β)

$$\lambda_f = \frac{1}{A\beta\left[\frac{1}{L_1} + \frac{1}{L_2}\right]}$$

where L_1 and L_2 are the effective fluid column lengths on each side of the piston

Compliance will reach a maximum value when $L_1 = L_2$, i.e. at mid-stroke, when

$$\lambda_f = \frac{L}{4A\beta}$$

This is the equation normally used for calculating fluid compliance since the mid-stroke position is normally the one at which stability conditions are at their worst. If the total fluid volume is to be taken into account, then the formula is modified to:–

$$\lambda_f = \frac{L}{4A\beta} + \frac{V_p}{4A^2\beta}$$

where V_p = total volume of fluid in the lines to the actuator.

Fluid compliance of a symmetrical linear actuator (Fig.98) compliance at any stroke position being defined as the ratio of the compliance at that position to the maximum compliance (at mid-stroke).

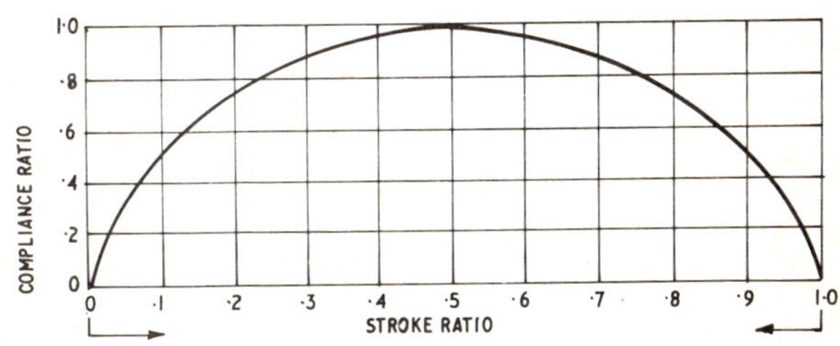

Fig. 98

Where F is the output force, the destabilising effect of compliance is equal to $- F\lambda$.

The flow is determined by the valve equation:

$$q = kwy\sqrt{P_o - \Delta P_L}$$

Since the actuator acts as an integrator in a position control system the output position (θ) can be expressed directly in terms of the flow, jack area (A) and effective compliance:–

$$\theta = \frac{1}{A}\int q\,dt - F\lambda$$

Essentially the servo is non-linear in characteristics (except for very small displacements).

Transfer function

The transfer function is a differential equation derived as defining the dynamic behaviour of a valve, component, or complete system.

The equation for output position derived above can be expressed in the form:–

$$\theta = \frac{K_V}{A} \int qdt \left(y + \frac{F}{S_V A} \right) - F\lambda$$

where K_V = value gain
S_V = valve output stiffness

Corresponding loop gain (K) $= \frac{K_V}{A}$

System output stiffness (S) $= - S_V A$

$$\text{where } \theta = \frac{K}{A} \int qdt \left[y - \frac{F}{S} \right] - F\lambda$$

For a pure inertia load the differential equation of motion is:

$F = Ms^2\theta$

where s = the Laplace operator for the function $\int qdt$

The open loop transfer function of the system then becomes:–

$$KG(s) = \frac{K}{S\left[\lambda ms^2 + \frac{Kms}{s+1}\right]}$$

where G(s) is the time dependent portion of the transfer function.

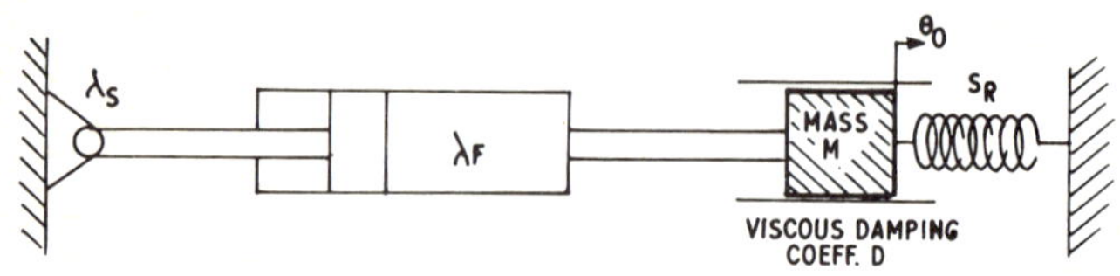

Fig. 99

For a system comprising inertia loading, a spring face and viscous damping (Fig.99) the differential equation of motion is

$F = ms^2\theta + S_r\theta + Ds\,\theta$

where D is the coefficient of viscous damping
S_r is the spring rate

The corresponding open-loop transfer function for the system is:–

$$KG(s) = \frac{K}{\lambda ms^3 + (\lambda D + \frac{Km}{s})s^2 + (\lambda S_r + \frac{KD}{(s+1)})s + \frac{KS_r}{s}}$$

The closed loop transfer function for a corresponding system can be rendered as:

$$\frac{\theta_O}{\theta_L} = \frac{KG(s)}{1 + KG(s)}$$

where θ_O = output movement or displacement

θ_L = input movement or displacement.

(See also chapter on Valve Characteristics)

Analysis on a mathematical basis requires specialised knowledge and experience of the subject and can be extremely tedious and time-consuming (in the absence of computers) although suitable solutions may be derived graphically. It is also impossible to generalise about the response which will be obtained with various systems as the controlling parameters may vary appreciably. It may be said, however, that the overall stability of the system will determine the amount of amplification (or amplifier gain) which can be used to increase the error signal level in closed loop systems.

The effect of increasing gain is to improve the response at the expense of the stability margin, and a high gain system with appreciable inertia will tend to overshoot during a transient before settling down at the final position signalled by the input. This is not necessarily a bad thing as limited overshoot characteristics may be perfectly acceptable as long as a steady final position is realised within a suitable time interval. On the other hand, with excess gain the overshoot characteristics may be greater than the damping present so that the system becomes oscillatory and useless for control purposes.

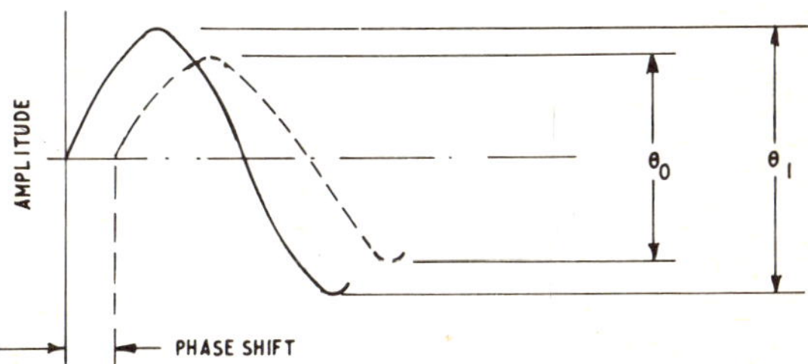

Fig. 100

Frequency Response

The frequency characteristics of any element in a loop can readily be determined with respect to a sinusoidal input, in terms of the difference between the input and output amplitudes, and the phase shift – Fig100 Such data can be determined empirically, thus expressing the frequency characteristics of a component; or derived mathematically or graphically) from the transfer function. The latter method leads to numerous different methods of deriving solutions for the differential equations involved. One such method in widespread use is to represent the transfer function logarithmically, introducing the decibel as the logarithmic unit for the modulus.

POWER SOURCES – ACTUATORS, AND FILTRATION

The choice of power source for hydraulic servo valves depends principally on the following–:

1. Valve input pressure
2. Maximum flow
3. Duration of maximum flow
   (a) % of total time
   ((b) longest continual demand
4. Estimated horsepower

First cost is kept low if a popular gear or vane pump is used, the pressure being regulated by passing surplus fluid through the relief valve (Fig.101)This gives a constant pressure supply but may involve cooling problems if the power exceeds about 5 horsepower.

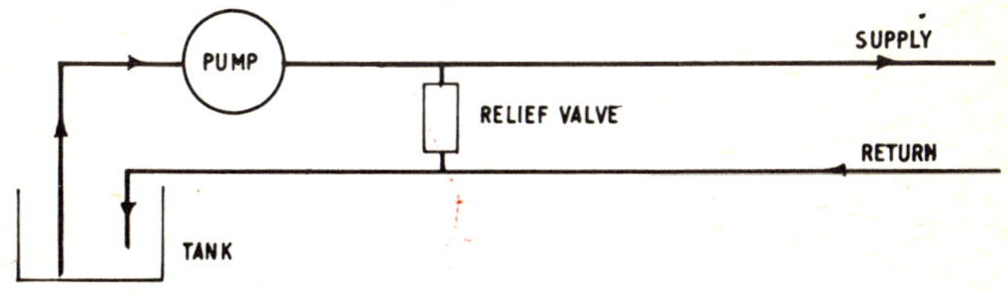

Fig. 101.

As an approximate guide, gear pumps are generally suitable for pressures up to 1,000 psi; and vane pumps for pressures up to 1500–2000 psi. For higher pressures a piston pump would be the normal choice Piston pumps are also used for lower pressures where the demand (maximum flow) is relatively high.

Many servo valves work intermittently and maximum flow is required for very short periods. A simple powerpack working in conjunction with an accumulator may then prove an ideal arrangement. The accumulator is charged during idle periods and the pump is unloaded until the pressure falls. (Fig. 102)

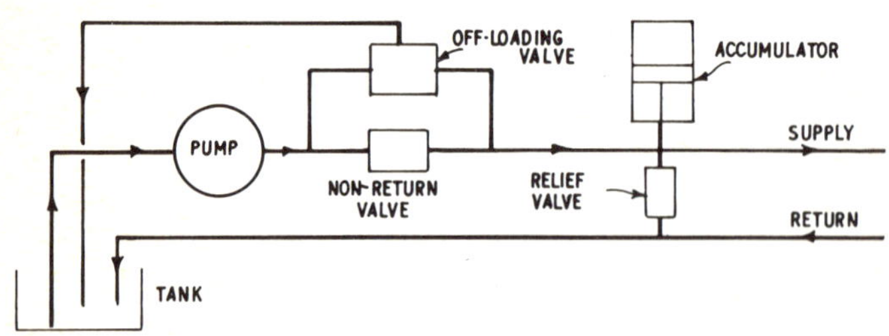

Fig. 102.

Some pressure fluctuations are unavoidable with this system if the full use is to be made of the accumulator capacity. A gas loaded accumulator can only release energy if the pressure is allowed to fall. If, on the other hand, it is only used to sustain pressure during idle periods and the pump is large enough to deal with the working load, then the only pressure variation would be that dictated by the pump unloading valve. The accumulator can then be comparatively small.

An accumulator may also be employed to eliminate pump pulsations, which may be objectionable in the case of high pressure piston pumps where pressure surges are usually apparent. In this case an accumulator can provide effective pulsation damping.

The minimum size of accumulator needed to reduce pump pulsations to an acceptable degree can be estimated from the following empirical formula:—

$$\text{Accumulator volume (gallons)} = \frac{C_p \times \text{pump discharge (gallons per min)}}{100}$$

where C_p is a constant depending on the type of pump, viz

= 5.0 for a single-cylinder single-acting piston pump
= 2.5 for a single-cylinder, double acting piston pump
= 0.45 for a two-cylinder, double-acting piston pump

The accumulator size given by the above formula is a minimum for effective damping, for pump speeds above 100 rpm.

The effect of pressure variations on valve flow must be considered, as this might be reduced at a vital moment, so affecting the behaviour of the actuator.

Pressure controlled variable delivery pumps are both efficient and suitable for any commercial pressure. The usual method is to operate the pump stroking mechanism with a spring opposed cylinder (Fig103) The rise in pressure as demand falls off causes the flow to fall to that sufficient for leakage. The difference in pressure between full and no flow depends on the spring rating. It can usually be assumed that the pressure available at full flow is 90% that at no flow,

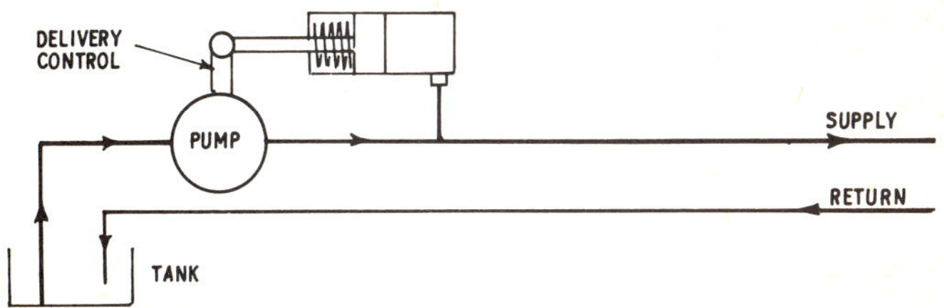

Fig. 103.

although the actual pressure can fall below this if the maximum pump flow is not sufficient to meet the demand. Typical pressure/flow characteristics are shown in Fig.104)The maximum power required to drive the pump is that for maximum pressure and flow, the maximum pressure depending on the application. Full pressure may only be required at partial flow and a suitable spring would ensure that the motor would not be overloaded at full flow.

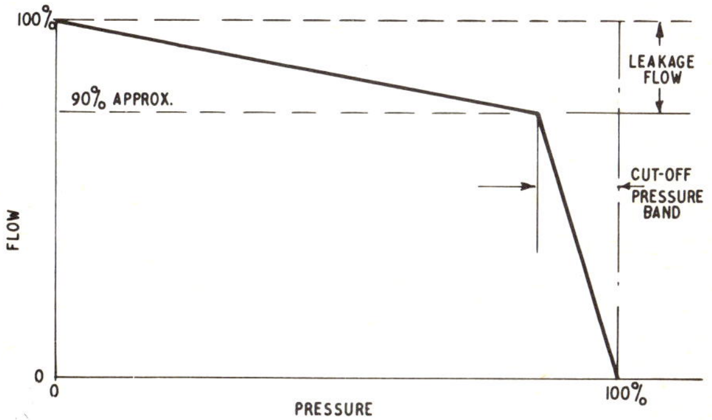

Fig. 104.

For pressures up to 1000 psi simple variable displacement pumps are available; but for higher pressures, plunger pumps, usually of the axial type, are essential.

PRESSURE AND FLOW

For industrial systems pressures within the range 500 to 1500 psi are in common use. Higher system pressures of the order of 2000 to 4000 psi are usual in aircraft hydraulic controls and has also influenced the choice of pressures for industrial servo systems. Thus the latter are tending to adopt somewhat higher pressures, typically 2000 psi, with a potential reduction in the bulk and weight of actuators required to produce a given output. The main component affected is, however, the pump. At the same time, as far as servo systems are concerned any gain in the reduction in bulk of components with increasing system pressure is offset by higher strength requirements and the necessity for more precise manufacture; and the increasing elasticity which may be introduced in the system (particularly the compressibility of the fluid). For industrial hydraulic servo systems, therefore, system pressures are generally limited to moderately high pressures (1500 to 2000 psi).

With any system two distinct pressure levels are involved – the steady state pressure associated with normal driving of the actuator, and the dynamic pressure level consistent with accelerating the actuator load. The latter represents an additional pressure loss over and above the 'steady state' losses or pressure drops in the actuator and system lines. The actual working pressure levels are thus determined as:–

Steady state pressure (normal driving) $= P - \Delta P_1 - \Delta P_a$

where P = pump delivery pressure

ΔP_1 = pressure drop in lines

ΔP_a = pressure drop in actuator

Pressure drop in the lines (and associated components) can be controlled by line sizing and is commonly based, on an 'optimum maximum loss' basis – e.g. a total value for P_1 of not more than 5% of the nominal system pressure (P). This steady state value may, however, be modified by dynamic considerations, e.g. accelerations modifying flow rates.

The steady state pressure drop in linear actuators can be derived directly from

$$\Delta P_a = \frac{F}{A\eta_a}$$

where F = linear output force

A = piston area

η_a = actuator efficiency

(the above in consistent units)

In the case of a rotary actuator the corresponding formula is:–

$$P_m = \frac{24\pi T}{D_m \eta_m}$$

where T is the motor torque

D_m = displacement of motor per revolution

η_m = motor efficiency

Where acceleration of the actuator is involved, then an additional load (or pressure) is imposed on the actuator which can be calculated on the basis of the modified force output, viz –

For a linear actuator:–

For a linear actuator:–

Additional pressure drop $(\Delta P_d) = \dfrac{Mf}{A\eta_a}$

where M = mass of moving parts

f = acceleration

(in consistent units)

For a hydraulic motor:

$$\Delta P_d = \frac{24\pi If}{D_m \eta_m}$$

where I = polar moment of inertia

Thus:

Dynamic state pressure (with acceleration) $= P - \Delta P_1 - \Delta P_a - \Delta P_d$

A further additional load (and effective pressure drop) may be involved in systems which have a high starting resistance or 'break-out' force.

Delivery flow is related directly to the pump displacement, speed and efficiency:–

$$Q = \frac{q_p n_p \eta_p}{277.4} \text{ gallons per minute}$$

where q_p = pump displacement, cu.in per rev

n = pump speed, rpm

Flow or delivery required by a linear actuator is:–

$$Q_a = \frac{A\ V}{277.4\ \eta_a} \quad \text{gallons per minute}$$

where V = linear speed (inches/minute where A is in square inches)

Flow or delivery required by a hydraulic motor is:–

$$Q_m = \frac{D_m \times rpm}{277.4\ \eta_m}$$

From the above can be determined the pump power rating and duty cycle. Pump performance is normally defined by pressure, delivery and power, although it does not automatically follow that a particular pump is rated for continuous operation at full pressure and flow (Fig.5). Instead the pump may be rated for constant power only (Fig. 6). In this case it may be capable of accepting transient overloads within the balance of the 'corner horsepower' envelope. Consideration of the complete duty cycle will indicate satisfactory, or optimum, pump characteristics. Thus with fluctuating power requirements it may be more economic to select a pump with a lower constant power rating (Fig.105) than needed for transient conditions, provided the latter yield acceptable transient overload(s). Such transients may only be associated with starting, accelerations and dynamic braking. Analysis of the duty cycle will also indicate whether or not the corner horsepower is exceeded at any point, which could set a demand for power in excess of the input power available from the pump alone.

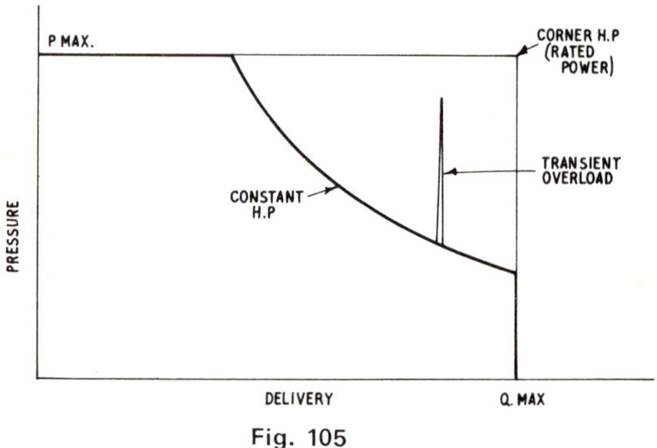

Fig. 105

ACTUATORS

Cylinders or jacks

These may be either single or double rodded, the advantage of the latter being that the characteristics are the same in both directions and the flow through the valve is symetrical in both directions. It will be appreciated that a servo valve can control the piston movement by acceleration, constant velocity, braking and reversal.

The fluid within the jack, being compressible, acts as a spring and can influence the behaviour considerably by reducing the rigidity of the system. Pressures below 500 psi or above 3000 psi may accentuate this tendency and movements controlled by jacks are usually restricted in length because of this. It can be assumed that the worst condition is at midstroke, and not as might be thought, at the end. The reason for this is that the valve controls both the incoming and exhausting fluid, and that once past mid-stroke the exhaust end contains less fluid and is therefore less 'springy'. This subject is dealt with in some detail in the chapter on System Stability.

Hydraulic motors

A hydraulic motor is usually similar to a pump made by the same makers, but modified if necessary to ensure that full pressure can be applied to both ports. It may also be necessary to provide for lubrication when starting under full pressure, as distinct from a pump which usually starts from zero pressure.

A motor, together with its driving gear, rack and pinion lead screw etc., will normally be appreciably more expensive than a jack. Its advantage lies in its small inertia and greater rigidity, giving a more positive action and one less influenced by any disturbing force.

Hydraulic motors with a servo valve incorporated are now widely used for driving the slides of numerically controlled machine tools, usually through lead screws with recirculating ball or hydraulic nuts, both giving precision combined with low friction. The relative inefficiency of a system where the hydraulic motor is controlled directly by the servo valve restricts this method to comparatively low powers, or where convenience is of greater value than thermal efficiency.

Large hydrostatic drives

Large hydraulic motors and those intended for operating continuously as part of a speed controlled drive, are usually coupled to a servo-controlled pump and drives up to 250 hp are possible with this arrangement. The servo may be programmed or controlled by the output speed.

Two methods of coupling are available with single line and reservoir. (i) when the drive is always in one direction and braking is not required; and (ii) close coupled which provides for dynamic braking and reversal. Most schemes of this kind have a boost pump which makes up for leakage and may circulate oil round the motors for cooling and then through a cooler and filter, A seperate system with its own pump reservoir and filtration may be needed for the servomechanism.

Where precision control is required any backlash can be the cause of relatively serious errors. High speed motors usually require reduction gears which may have to be specially designed to eliminate backlash.

It is also possible to do this hydraulically, one method employing twin motors to drive a rack. The motors are controlled by a single valve and this causes the opposing gears to 'grip' the rack.

Another method employs helical rack with a pair of gears, one being thrust into the rack by a small cylinder until the backlash has been taken up.

Semi motors or rotary actuators.

Some types of process control valve such as plug and butterfly valves require only a partial turn for opening and shutting. These and similar applications can be catered for by semi motors or rotary actuators.

A very compact arrangement is that shown in Fig. 106 where a rotary vane is contained within a circular housing. With the single vane shown, the maximum angle of rotation is about 300° but by having a double vane, with two inlets and outlets, the power for a given size can be doubled, whilst reducing the angle of rotation to about 100°. This is, of course, ample for the valves mentioned.

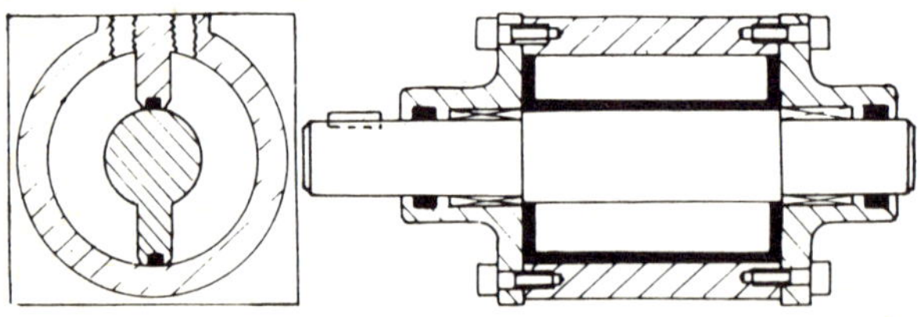

Fig. 106.

Another type of semi-rotary actuator embodies a rack and pinion, the rack being actuated by one or more cylinders. A common construction (Fig.107) is where there are two racks, each with its own single acting cylinder. This has the effect of eliminating backlash in the same way.

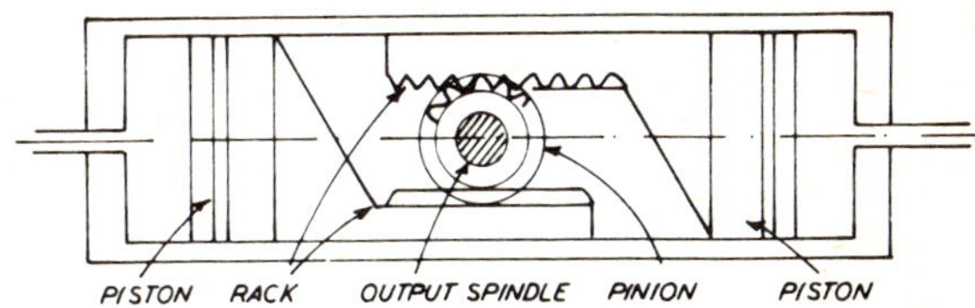

Fig. 107.

Fluids

The majority of hydraulic servo-systems employ mineral oils, either those specially blended for hydraulic systems or for steam turbines. Unless seriously overheated these oils have an indefinite life and the only treatment required is provided by the normal filtration recommended for the removal of solid particles.

Fire resistant fluids

With the ever increasing number of servo systems there will be some which are installed in situations where it is considered that mineral oil presents an unacceptable fire hazard.

There are a number of commercial fire resistant fluids available, the principle varieties being –

Water based fluids
Water/glycol (about 35% water)
Water/soluble oil (about 95% water)
Synthetic Fluids
Phosphor ester base
Chlorinated diphenyl base

Water based fluids are the least expensive and have a high bulk modulus, giving greater rigidity to the system. Their use in servo systems would, however, be only considered in extreme cases.

Synthetic fluids are more like mineral oils in many respects, but they cost at least six times as much and also involve extra expense in equipment and maintenance.

Synthetic fire resisting fluids are compatible with most metals but seal materials require special consideration. Accumulator bags are particularly vulnerable.

Unlike mineral oils, fire resistant fluids cannot be left to look after themselves. The first obvious difference is their lack of a stable viscosity and the viscosity falls rapidly after pumping through narrow passages such as those which often occur in servo systems. The change of viscosity may vary considerably with fluids obtained from different suppliers.

Synthetic fluids lack the chemical stability of mineral oils and are influenced by moisture and contact with the air. It is common practice to filter a portion of the flow through Fuller's earth to remove any tendency to acid build up, but displaced particles of Fuller's earth can cause a mechanical filtration problem.

These fluids, unlike mineral oil, are heavier than water and any moisture tends to rise to the surtace. Use is made of this fact to skim off a small proportion and pass it through a high vacuum chamber where most of the water flashes off and is discharged.

Any user of fire resistant fluid must be prepared to make arrangements for checking the fluid condition at regular intervals and to make sure that any service personnel fully understand the importance of the adverse effects of moisture.

FILTRATION

The trouble free operation of most hydraulic servo systems depends on fluid cleanliness. Electro-hydraulic valves particularly tend to have very small passages, particularly when near the null position, which can act as filters themselves and quickly silt up if the fluid is contaminated. Metallic particles, if jammed between spool and housing may cause physical damage.

It can always be assumed that a new hydraulic system is dirty by these standards and the system should be run for at least twelve hours with filters in position but with the valve itself replaced by a special block. The filter element is then replaced by a new one and the system run for another hour or so before fitting the servo valve.

When a filter element has to be changed during service the servo valve should again be replaced by the block whilst the new filter is run for a short time. Changing a dirty filter is bound to cause some of the trapped contamination to be dislodged and this must be removed before it can do any harm.

A 5-micron filter should be fitted in the pressure line as near the valve as possible and some users also insert a magnetic filter which will catch ferrous particles too small to be caught by the main filter.

In a high pressure micro filter specially developed for servo valves (Fig. 108), the actual filter element is of sintered bronze and will withstand a considerable pressure without damage should the flow be restricted. When clogged the element is removed and may be cleaned in an ultrasonic bath.

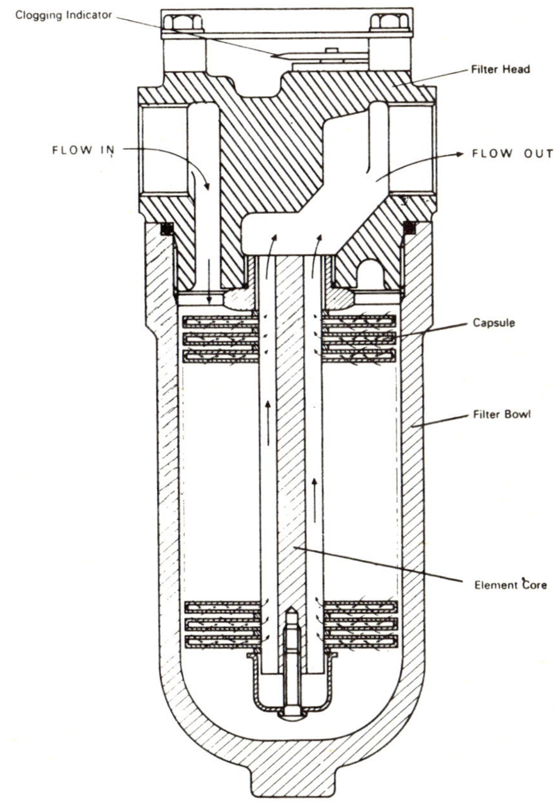

Fig. 108

Fairey high pressure micro-filter (sintered bronze capsule element).

A combined magnetic and mechanical high pressure filter is shown in Fig.109. It will be noted that the magnet is downstream of the replaceable filter element, as it has been found advisable to restrict the magnet 'catch' and use it to trap only those ferrous particles which have passed the element.

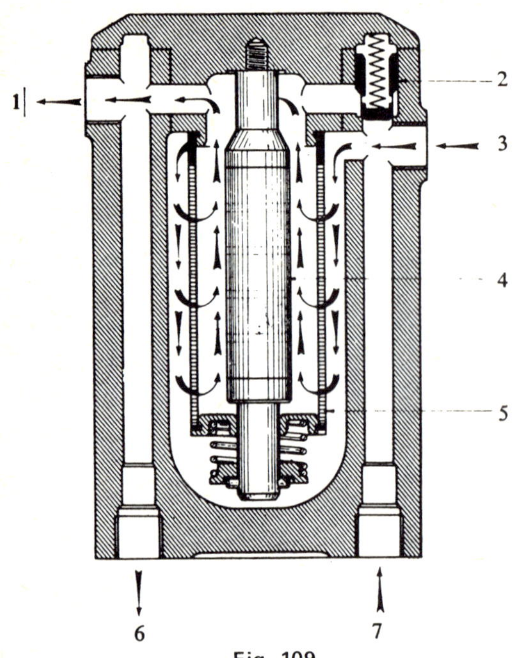

Fig. 109
High pressure micro-filter with magnetic and mechanical elements (Pratt Precision Hydraulics Ltd.)
1. Side outlet. 2. Relief-valve. 3. Side inlet. 4. Magnetic filter element. 5. Replaceable micro-filter element. 6. Alternative bottom outlet. 7. Alternative bottom inlet.

One of the main sources of contamination is atmospheric dirt and a fine air filter on the reservoir breather is essential. Make-up fluid should be treated as if it were dirty and if possible should be pumped from the container to the reservoir through a 5-micron filter fitted permanently for the purpose. This will also avoid having to open the reservoir to the atmosphere.

Completely sealed reservoirs are usually impracticable. Pressurized reservoirs have been used but they have the disadvantages that they tend to increase the amount of air dissolved in the fluid and this may affect the system behaviour.

Pumps and actuators generate contamination continuously and high pressure oil can cause errosion of material from pipes and fittings. This is why the main filter must be put on the high pressure side of the valve and as close to it as possible.

A hydraulic system should never be broken into unnecessarily as this is the surest way to introduce contamination and start leaks by disturbing seals which had settled down. Filter dondition can be judged by the rise in differential pressure across it as it silts up. It should be remembered, however, that the pressure does not show any significant increase until the need for renewal is imminent. For this kind of application it is doubtful if a coarse filter can serve any useful purpose in prolonging the life of the fine filter.

As internally generated contamination is chiefly ferrous, magnet probes are sometimes inserted in the system in special adaptors which enable them to be withdrawn under pressure. Periodic examination of the probe can give advance warning of unusual wear in a major component. It does not, however, detect wear or disintegration of elastomeric parts, such as accumulator bags and 0-rings.

INDEX